Systems Analysis and Operations Management

Richard J. Hopeman

Syracuse University

Charles E. Merrill Publishing Co.

A Bell & Howell Company
Columbus, Ohio

Library of Congress No.: 69-19269

Standard Book No.: 675-09514-X

1 2 3 4 5 6 7 8 9 / 76 75 74 73 72 71 70 69

Printed in the United States of America

*to Lori
and Mark*

Preface

In an age of computers and space exploration, we are confronted with exciting breakthroughs in knowledge which radically change our perspectives. Inherent in many of these bold forward strides is the application of the systems concept. This book explores the possibilities of the use of that concept in the field of operations management.

Part I of the book is devoted to general concepts in the systems area, to the environmental set of the firm, to the firm as a system, and to the design of systems. The ideas presented are applicable to all types of systems within the firm.

In Chapter 1, "Systems Design," the challenges and opportunities of the systems concept are explored. A systemic hierarchy is presented which provides a framework for the organization of the book. Consideration is given to the fundamental nature of systems and the analysis of flow patterns.

Chapter 2, "Systems Analysis," explores the systems concept and decision making, various types of models useful in decision systems, and the use of simulation in system design, systems analysis, and decision making. A procedure for the development of systems is proposed and illustrated with an example which emphasizes systems conceptualization, model building, and simulation.

In Chapter 3, "The Systems Concept," the fundamental ideas involved in the concept are investigated. Since many of these ideas challenge the traditional approach to management, special attention is given to the successful application of the systems concept in other fields to illustrate the merit of this alternative approach. These fields include astronomy, biology, chemistry, and medicine. The contrasts between the breakthroughs in knowledge in these fields and the field of business bring into sharp focus the valuable potential contribution the systems concept can make in advancing management theory.

Chapter 4, "The Environmental Systemic Construct," explores the relationships of the firm to those groups in its environment with which it interacts. These interaction patterns are traced with respect to customers, competitive firms, communities, labor unions, stockholders, banks, suppliers, and governments. In each case, a systems approach is used which raises fundamental questions about these relationships. Several new systemic patterns are proposed which may alter significantly the practice of management.

Chapter 5, "Organization Theory: Classical, Neoclassical, and Systems," contrasts the systems approach and two other commonly encountered approaches to management. In order to visualize the role of a systems approach within the firm, certain management concepts are explored, their deficiencies are explained, and the use of a systems approach to overcome these deficiencies is discussed.

Following an exploration of the environment of the firm and evaluation of traditional concepts of management within the firm, "The Firm as a System" is investigated in Chapter 6. The systems approach provides a dramatic contrast and alternative to these traditional concepts. The systems approach is developed in terms of the planning and control of four basic flow networks which permeate the firm. The organizational consequences of a systems approach, which significantly modify current concepts of organization structure, are then developed.

In Part II, in order to analyze the concepts in depth, the systemic scope is restricted to system design and systems analysis as applied to operations management of the materials flow network. This represents one of the key flow networks of the firm.

The materials flow network is composed of four management information subsystems. The inventory planning and control subsystem is developed in Chapters 7, 8, and 9. The logistics subsystem is discussed in Chapter 10. The inventory stock status subsystem and the purchasing subsystem are examined in Chapter 11. In each of these four subsystems, a variety of decision models are explored. They include least squares regression, moving averages, exponential smoothing, forecasting techniques for cyclical fluctuations, economic order quantity models, inventory turnover models, Monte Carlo simulation, queuing theory, and linear programming. Although these models are important at decision points within the management information system, special emphasis is given to the dynamics of the information system for planning and control. The information systems are designed to utilize computers.

In Chapter 8, "Real-Time Systems," the potential is explored for incorporating the ideas presented earlier in the book in an immediate-response management information system. Since this advanced concept is not now fully developed in business firms, several applications are drawn from other systems to illustrate how the concept has been applied successfully in practice. These applications include telephone systems, electrical systems, aerospace systems, airline operations systems, and the National Airspace Utilization System. A review of these systems indicates the remarkable potential for the application of systems analysis in operations management.

In the rapidly evolving field of systems analysis in business management, several authors have taken widely divergent approaches. This is typical of the evolution of new concepts. This book is no exception. Many new and controversial ideas are explored and proposed. In a sense, more questions are raised by these ideas than are answered. For this reason, an effort has been made to develop a comprehensive bibliography. The multitude of sources cited should provide sufficient material for those wishing to do research in the systems area or develop systemic solutions to operations management problems.

Richard J. Hopeman

Contents

12   Real-Time Systems   293

Bibliography   305

Index   341

Part I

Systems Analysis

Systems Design

The Challenges and Opportunities of the Systems Concept

1

The present age is one of rapid technological change. In conjunction with this change is the development of a high degree of specialization by the people in organizations. These people develop advanced skills in depth, but do so in ever narrowing fields of specialization. As organizations have acquired more specialists, they also have grown to emerge as massive, complex enterprises. Together, these developments have led to a vast array of new products and services, many of which had hardly been imagined a few years ago. At the same time, these developments and trends have created unprecedented managerial challenges as well.

The management of large-scale operations, faced with a multitude of technological changes and staffed by highly competent specialists, requires, above all else, skill in integration and synthesis. Most of all, effective management of such operations requires skilled generalists rather than a collection of functional specialists: executives who can plan, analyze, and control complex operations in an integrated manner.

Important advances have been made by theorists and practitioners in management science and organizational behavior. Although these advances have been directed at the management of complex organizations, they often have been limited to solving specific problems in the firm.

The significant challenge facing management remains that of

integration and synthesis. Traditional management approaches have failed to meet that challenge. The systems concept meets it head on. The success of the approach and its impact on management can be seen in a few innovative firms as well as in the Department of Defense under former Secretary McNamara and in the National Aeronautics and Space Administration in the management of its massive and complex Apollo program.

The systems approach is a significant departure from traditional management. Rather than being concerned with meeting the objectives of a narrow functional specialty, management through systems is concerned with meeting the overall objectives of the organization. This is accomplished through the use of a decision system and an information system which reflect the operation of physical systems or flow networks within the firm.

Flow networks can be classified in many ways. One such classification, which is developed later in the book, includes 1) the materials flow network, 2) the money flow network, 3) the manpower flow network, and 4) the machine, facility, and energy flow network. Inherent in the planning, analysis, and control of the inputs, transformation processes, and outputs of these networks is consideration of the impact of systemic changes in one network as it affects the operation of other networks.

Beyond the physical flow networks, the manager of systems must consider the impact of environmental changes from such sources as government, customers, competitive firms, the community, labor unions, stockholders, banks, and suppliers. His concern must be not only with how they affect the firm and how the firm affects them, but also with how they interact among themselves and with the firm in bringing about changes in the state of the business environment.

Thus, the systems concept integrates at the flow network level within the firm, at the level of the firm as a totality, and at the environmental level. It provides a vehicle for understanding and for the managerial planning, analysis, and control which achieve synthesis. The opportunity for its adoption in industry is great. The increased use of computers has fostered its development since careful system design, systems analysis, and programming are requisites for successful and profitable computer operations. At a very pragmatic level, the opportunities in the system area can be sensed by anyone simply by checking the vast number of openings in the field which appear regularly in the classified section of most metropolitan newspapers. Also of interest are the very attractive salaries associated with these positions.

If the opportunities are great, by comparison the challenges are even greater. The systems concept can be utilized at a variety of levels in an organization, from the design and analysis of an automated process to the development of a management system for the entire firm which incorporates environmental impacts. Although we have well developed hard data and reasonably solid systems engineering knowledge at the automation or process control level, we are faced with a greater challenge in the broader area of general management systems for the entire firm. Through the development of systems concepts for the firm and the practical testing of these concepts by innovative managers in recent years, more knowledge is being gained of how to use the concept effectively for general management.[1]

The greatest challenge lies in the development of a systemic understanding of the environment of the firm and the systemic impacts which are a part of it. Much research must be done in this area before the systems concept can provide its full benefit in operations management. Although the study of the environment is a significant challenge which may require years to complete, the systems concept *per se* provides a useful viewpoint and methodology for achieving such understanding.

Meeting the Challenge

Challenges to management such as those mentioned above are faced in this book by examining systems, the scope of which begins

---

[1] For insights into the management controversy over the utilization of the systems concept see the following articles: H. I. Ansoff, "The Firm of the Future," *Harvard Business Review,* September-October 1965; W. M. Brooker, "The Total Systems Myth," *Systems and Procedures Journal,* July-August, 1965; J. A. Brown, "What New Systems Must Do," *Administrative Management,* January, 1964; M. Carasso, "Total Systems," *Systems and Procedures Journal,* November, 1959; W. L. Cisler, "Management's View of the Systems Function," *Systems and Procedures Journal,* July-August, 1965; J. W. Culliton, "Age of Synthesis," *Harvard Business Review,* September-October, 1962; J. Deardon, "Can Management Information Be Automated?" *Harvard Business Review,* March-April, 1964; P. F. Drucker, "Long-Range Planning, Challenge to Management Science," *Management Science,* April, 1959; A. Harvey, "Systems Can Too Be Practical," *Business Horizons,* Summer, 1964; F. E. Kast and J. E. Rosenzweig, "Systems Theory and Management," *Management Science,* January, 1964; H. J. Leavitt and T. L. Whisler, "Management in the 1980's," *Harvard Business Review,* November-December, 1958; R. L. Martino, "The Development and Installation of a Total Management System," *Data Processing for Management,* April, 1963; W. G. Scott, "Organization Theory — An Overview and Appraisal," *Journal of the Academy of Management,* April, 1961; A. T. Spaulding, Jr., "Is the Total System Concept Practical?" *Systems and Procedures Journal,* January-February, 1964; S. Tilles, "The Manager's Job — A Systems Approach," *Harvard Business Review,* January-February, 1963.

at the broadest level and proceeds to narrower levels. This approach is depicted in Figure 1-1.

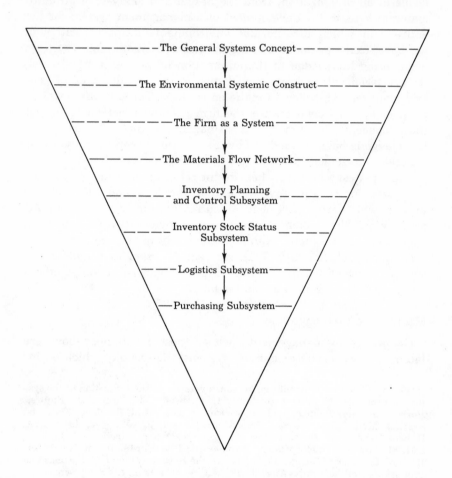

Figure 1-1.   A Systemic Hierarchy

At the primary level — the development of the general systems concept — the thrust is in the direction of understanding the universality of the systems concept as it is used in many fields. Although the systems concept could be treated in depth in a comprehensive analysis of its contribution to breakthroughs in knowledge in the physical sciences, social sciences, and applied sciences such as engineering, such a treatment would constitute a book in itself.[2] Our purpose in this book is served by developing some understanding of the general systems concept and reviewing comparatively a few selected major breakthroughs in the fields of astronomy, biology, chemistry, and medicine. With these ideas and demonstrated achievements of the systems concept in mind, then, we will consider next the environmental systemic construct.

At the environmental level we are concerned with developing an understanding of the impacts of the groups which interface with the firm but are external to it. We shall explore the inputs, transformation processes, and outputs which flow between the firm and the objects in the environmental set. The environmental set includes the objects depicted in Figure 1-2. In later chapters the relationship of

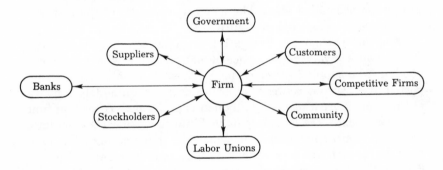

Figure 1-2. The Environmental Set of the Firm

each object to the firm will be explored, systemic and non-systemic patterns of interaction discussed and propositions for modified relationships presented. Although broad in treatment, the examination of the environmental systemic construct indicates the promise of a truly integrated management system and raises questions, which, it is hoped, will lead to further research as well as innovative experimentation in business firms.

---

[2] For further study of general systems theory see the following, as well as a variety of articles in *General Systems*. By L. von Bertalanffy, "An Outline of General Systems Theory," *British Journal of Philosophical Sciences* (1950); "General Systems Theory — A Critical Review," *Yearbook of the Society for General Systems Research*, Vol. 7 (1962); "General System Theory," *General Systems*, Vol. 1 (1956); and "General System Theory—A New Approach to the Unity of Science," *Human Biology*, December, 1951. By K. E. Boulding, "General Systems as a Point of View," in M. D. Mesarovic, *Views on General Systems Theory* (New York: John Wiley & Sons, Inc., 1964); and "General Systems Theory — The Skeleton of Science," *General Systems*, Vol. 1 (1956).

Having determined the limitations of traditional management theory in meeting the challenges now faced by managers of large, complex, technologically oriented firms we will focus our attention on the firm as a system. It is in this area that the flow networks concept is presented in some detail. Viewing the firm as a series of flow networks raises many new and controversial management theory questions but it also provides some new answers to old problems — problems that are growing acute as present trends continue to develop with respect to growth, specialization, and rapid technological change. After a consideration of the firm as a system, the focus narrows once more to encompass the materials flow network. This represents just one of the four flow networks in the systemic model of the firm, yet it provides a focus for the application of the systems concept at a detailed level. The general schematic of the materials flow network is depicted in Figure 1-3.

From the schematic in Figure 1-3 it can be seen that the materials flow network involves a sequence of materials flow which starts with vendors who supply inventory items. These items, after being transported over various routes via various carriers, ultimately converge at the firm where a receiving function is performed. After that function is exercised, the items are moved into raw materials inventories and other source inventories to which the items pertain. The items then go through patterns of convergence and divergence within the firm as they undergo transformation processes which yield finished products. Inherent in the transformation process is materials handling and a variety of in-process inventories. From the finished goods inventory, the products undergo a shipping function and diverge through channels of distribution to customers. These channels may be direct, involving only transit activities, or they may include channel inventories in the form of wholesalers and retailers.

In conjunction with the physical materials flow network are four decision and information subsystems which must be integrated for effective systems management. These include the inventory planning and control subsystem, the inventory stock status subsystem, the logistics subsystem, and the purchasing subsystem. These subsystems, the decision models used in them, and the information systems associated with them are examined in detail in Part II of the book. For introductory purposes here, they are described briefly below.

The *inventory planning and control subsystem* is the analytical center of the materials flow network. It is responsible for determining forecasts of inventory requirements, determining the appropriate

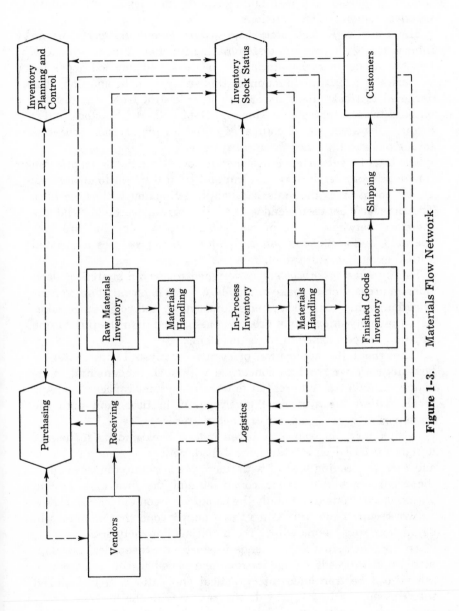

Figure 1-3. Materials Flow Network

quantities of items to stock and to order, determining when orders should be placed, and maintaining appropriate levels of inventory turnover, among other functions.

The *inventory stock status subsystem* essentially performs an information gathering and disseminating function. Tracking of stock levels of items in raw materials, in-process, and finished goods inventories is a critical function of this subsystem. In addition, it is designed to track inputs on order and in transit from vendors and outputs from shipping to customers. Data gathered on stock status in this subsystem are, of course, essential for the decision processes taking place in the other three subsystems.

The *logistics subsystem* is designed to coordinate the transportation functions from vendors to the firm and from the firm to customers. It also coordinates the materials handling function within the firm. The interface between vendors and the firm at receiving and the interface between the firm and customers at shipping are also operated under the decision and information processes associated with the logistics subsystem.

The *purchasing subsystem* is designed to handle analytical decisions in areas such as make-versus-buy, price determination, vendor selection, value analysis, systems contracting, and so forth. In addition, the many information processes associated with the purchasing function are performed in this subsystem.

Throughout the application of systems analysis to the materials flow network we shall be concerned with such fundamental interacting variables as differential flow rates, system capacity, volume levels within the system at given points in time, and with the dynamics of convergence-divergence patterns.

In summary, the approach taken in this book and its sequential design are based on starting with broad, general systemic concepts and then proceeding to deal with narrower systemic ranges such as the environmental systemic construct and the firm as a system within its environment. Finally the book focuses on the materials flow network within the firm as a system and beyond that examines in detail four subsystems within the materials flow network.

An approach such as this tends to start with theoretical developments and proceeds as the focus is sharpened to the point where theory can be translated into practical applications. It is believed that this approach will provide the reader with the necessary concepts and management models to meet the challenges confronting managers of complex organizations. In essence, it serves to integrate physical systems, management information systems, and decision systems toward the end of achieving organizational objectives through synthesis.

Since any new and emerging field — in this case, systems analysis — is characterized by a vocabulary of its own and by general operative concepts, the reader will find it helpful at this point in the book to become familiar with the vocabulary and some operative concepts of systems analysis.

The Nature of Systems

Systems analysis in business is still in an embryonic stage, and thus it is to be expected that the vocabulary of systems is growing rapidly and that total agreement on definitions still does not exist. For this reason we shall examine some aspects of the nature of systems and the vocabulary of systems as it is used in this book.

In order to examine the systems concept and utilize it in operations management, it is necessary first to examine carefully just what is meant by the term *system*. Many people in the field of systems analysis have developed their own definitions. Even though these definitions necessarily vary, enough similarity can be found to provide a workable basis for later analysis and synthesis using the systems concept.

Optner suggests that a system is "a set of objects with a given set of relationships between the objects and their attributes."[3] A closely related definition states that a system is "a set of objects together with relationships between the objects and between their attributes."[4]

Tilles states that "the basic notion of a system is simply that it is a set of interrelated parts."[5] Along the same line, Timms suggests that "a system is a set of elements so *interrelated* and *integrated* that the whole displays unique attributes. This definition implies that to understand the cause of the attributes of a system one must understand the interrelationship of the elements, called *subsystems*, and how they are *integrated*. If one were to cause a change in the interrelationships

[3] Stanford L. Optner, *Systems Analysis for Business and Industrial Problem Solving* (Englewood Cliffs, N.J: Prentice-Hall, Inc., 1965), p. 26.

[4] "The Definition of a System," *Yearbook for the Advancement of General Systems Theory* (1956), p. 18.

[5] Seymour Tilles, "The Manager's Job—A Systems Approach," *Harvard Business Review*, January-February, 1963, p. 74.

of the elements and thereby bring about a different integration, one would expect the system to display different attributes."[6]

Hall states that "a system is a set of objects with relationships between the objects and between their attributes.[7] Johnson, Kast, and Rosenzweig perceive a system as an "array of components designed to accomplish a particular object according to plan."[8] Similarly, McDonough and Garrett define a system as "a means for accomplishing some purpose or set of purposes. A full description of any system therefore requires the spelling out of (1) the specific expected accomplishments and (2) the specific mechanisms and procedures which are to be used in the processes."[9]

McMillan and Gonzalez point out that although we use terms such as inventory systems, distribution systems, production systems, and information systems commonly in business firms, it is useful for analytical purposes to consider a system as "a set of objects together with relationships between objects and between their attributes."[10]

Chin indicates that "the analytic model of system demands that we treat the phenomena and the concepts for organizing the phenomena as if there existed organization, interaction, interdependency, and integration of parts and elements. System analysis assumes structure and stability within some arbitrarily sliced and frozen time period."[11]

Ellis and Ludwig propose that "a system is a device, procedure, or scheme which behaves according to some description, its function being to operate on information and/or energy and/or matter in a time reference to yield information and/or energy and/or matter."[12]

---

[6] Howard L. Timms, *The Production Function in Business* (Homewood, Ill.: Richard D. Irwin, Inc., 1966), p. 101.

[7] Arthur D. Hall, *A Methodology for Systems Engineering* (Princeton, N.J.: D. Van Nostrand Co., Inc., 1962), p. 60.

[8] Richard A. Johnson, Fremont E. Kast, and James E. Rosenzweig, *The Theory and Management of Systems* (New York: McGraw-Hill Book Company, 1967), p. 113.

[9] Adrian M. McDonough and Leonard J. Garrett, *Management Systems: Working Concepts and Practices* (Homewood, Ill.: Richard D. Irwin, Inc., 1965), p. 2.

[10] Claude McMillan and Richard F. Gonzalez *Systems Analysis, A Computer Approach to Decision Models* (Homewood, Ill.: Richard D. Irwin, Inc., 1965), p. 1.

[11] Robert Chin, "The Utility of Systems Models and Developmental Models for Practitioners," in Schoderbek, *Management Systems: A Book of Readings* (New York:, John Wiley & Sons, Inc., 1967), p. 17.

[12] David O. Ellis and Fred J. Ludwig, *Systems Philosophy* (Englewood Cliffs, N.J.: Prentice-Hall, Inc., 1962), p. 3.

Neuschel suggests that "a system is a network of related procedures developed according to an integrated scheme for performing a major activity of the business."[13]

The most colorful definition of all is in Kenneth Boulding's inimitable poetry:

> A system is a big black box
> Of which we can't unlock the locks
> And all we can find out about
> Is what goes in and what comes out.
>
> Perceiving input-output pairs,
> Related by parameters,
> Permits us, sometimes, to relate
> An input, output, and a state.
>
> If this relation's good and stable
> Then to predict we may be able,
> But if this fails us — heaven forbid!
> We'll be compelled to force the lid! [14]

At this point one might be able to select a workable definition of *system*. Twelve current ones have been presented to the reader and by reviewing the extensive bibliography the reader could find several hundred more. Although it may seem likely that one definition of system will emerge in the field of systems analysis, the author doubts it since the systems concept fundamentally applies to many disciplines and cross-disciplines. Indeed, the systems concept will create future disciplines, each of which will define it in its own way — and even then there will be disputes.

Nevertheless, there are certain aspects of each of these definitions which provide some rough dimensions of what might be considered to be a system. Several of them mention objects, attributes, and relationships. The objects represent the components or elements in the system. Their attributes or characteristics provide measures of the objects as they change over time and as their systemic relationships are modified. The relationships, of course, provide the linkages among the objects and provide integration for the system. Optner provides even more formal definitions of these three terms.

---

[13] Richard F. Neuschel, *Management by System* (New York: McGraw Hill Book Company, Inc., 1960), p. 10.

[14] Mihajlo D. Mesarovic, *Views on General Systems Theory* (New York: John Wiley & Sons, Inc., 1964), p. 39.

Objects are the parameters of systems; the parameters of systems are input, process, output, feedback-control, and a restriction. Each system parameter may take a variety of values to describe a system state.

Attributes are the properties of object parameters. A property is the external manifestation of the way in which an object is known, observed, or introduced in a process. Attributes characterize the parameters of systems, making possible the assignment of a value and a dimensional description. The attributes of objects may be altered as a result of system operation.

Relationships are the bonds that link objects and attributes in the system process. Relationships are postulated among all system elements, among systems, and subsystems, and between two or more subsystems. Relationships may be characterized as first order, when they are functionally necessary to each other. Symbiosis, a first-order example, is the necessary relationship of dissimilar organisms; for example a plant and a parasite. Relationships may be characterized as second order if they are complementary, adding substantially to system performance when present, but not functionally essential. Synergy is a second-order relationship. Synergistic relationships are those where the cooperative action of independent agencies taken together produce total effects greater than the sums of their effects taken independently. Relationships may be characterized as third-order when they are either redundant or contradictory.

Redundance describes a state whereby the system contains superfluous objects.

A *contradictory* condition exists when the system contains two objects, which, if one is true, the other by definition is false.[15]

In applications to business decision making, such concepts as those above become particularly meaningful with respect to symbiosis and synergy. In practice, most business problems today are solved by a limited attack on the particular problem at hand. Typically its ramifications are not fully explored. Thus an expedient decision, perhaps the result of time pressures, may overlook the symbiotic relationships of one decision with others in the firm. What seems to be an optimal course of action to one manager may create problems for others and result in suboptimization for the firm. Conversely, an understanding of symbiotic relationships within the firm should yield decisions which are mutually reinforcing among managers of various areas of the firm.

---

[15] Optner, *op. cit.*, pp. 26-27.

The concept of *synergy* is particularly important in the business firm. In engineering systems it is common to measure system effectiveness in terms of the ratio of outputs over inputs. Due to friction and heat losses in the operation of engineered systems, this ratio typically yields something less than a ratio of one to one. In business firms, however, the value of the outputs must exceed the value of the inputs if the firm is to survive and grow. This measure of synergy is usually profit although we shall see later that it can be measured in other useful ways. Synergy in business systems is not just important — it is essential.

In addition to the concepts of objects, attributes, and relationships, which are common to many of the definitions cited, there appears to be some distinction between those who view systems with respect to procedures and those who view systems with respect to elements or objects. Reasonably enough, those who formerly were concerned with "systems and procedures" hold the former view. Their concern is with paperwork flow and particular office procedures such as payroll accounting, billing, and so forth. Representatives of the latter view are researchers who take a very broad theoretical approach to the inquiry into systemic relationships, i.e., the searchers for and builders of the "black box" to which Boulding refers.

Although there is a difference in the literature between these two approaches to systems, in time they will be merged. The systems theorists, if they have hopes of seeing their concepts tested operationally, must lay out the broad design for systems through analysis of objects, attributes, and relationships. Following that they will require the knowledge of the procedures group to refine the design and to block diagram in detail the logic involved. To reach the level of operational effectiveness, particularly on computers, requires procedural analysis and design down to the point of decimal placement — the computer equivalent of crossing all the t's and dotting all the i's.

A few of the definitions of the term *system* place some stress on the term *objective*. Business systems indeed must have objectives if they are to be judged useful and the nature of the objectives chosen has a significant impact on the design of the system. The perplexing question is, "What are the objectives?"

From a systems design point of view, it is easiest to design a system when the objectives are unambiguous and quantifiable. It also helps to know the mathematical relationships which will yield optimization of the objectives. Unfortunately, business problems cannot be solved with mathematical models alone. The variables and constants involved are usually a mixture of quantifiable and nonquantifiable factors. Even such common objectives as profit maximization are quite ambiguous when considered at the detailed system design level.

An even more perplexing problem associated with the determination of the objectives of a system is the situation in which traditional economic concepts of firm performance are abandoned in favor of the view that the firm is a social institution which serves social and psychological needs of its members. The inputs, transformation processes, and outputs in such a system are cast in the image of behavioral science. Although in such a behavioral construct the objects (individuals, groups, etc.) can be identified, their attributes are difficult to define and measure and their relationships are difficult to determine.

However one chooses to define the objectives of a system, it is important to recognize that before effective systems design can be started, some objectives must be determined. In this sense objectives, and the plans which stem from them, are inherent in the concept of a system.

The Flow Pattern of a System

As mentioned earlier in the chapter, a system can be generalized as a flow pattern. In this flow pattern inputs flow through a transformation process to become outputs. The inputs represent those objects in the system which initiate action with respect to the flow and the outputs represent objects which serve to meet the objectives for which the system is designed. In a business application, the transformation process should increase the value of the inputs as they are converted to outputs — the concept of synergism.

In the input-process-output construct, the critical problems are the determination of the desired outputs with their associated measurements, the determination of the inputs, and the routing, capacity, and flow rates through the transformation process as well as the development of the logic associated with the transformation process.

Associated with the input-process-output construct is the concept of feedback and control. Just as the design of a system implies objectives and plans, its successful operation implies controls. These controls essentially monitor systemic flows and through measurement of system status signal the need for action when the results being

achieved do not conform to the original plans. The feedback loop
concept implies that signals are returned from late stages in the
process (or in the outputs) to early stages in the process (or in the
inputs) to provide correction of the process. In sophisticated systems
this feedback loop may be automatic and capable of providing self-
correction and self-control as a characteristic of the system. Examples of this abound in automated equipment designs utilizing
servomechanisms.

The input-process-output flow patterns in given systems may take
many forms as they are linked together. At the subsystem level, for
example, three subsystem flow patterns may be sequential as indicated in Figure 1-4.

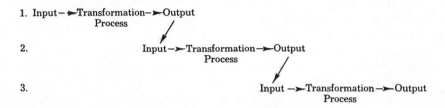

Figure 1-4.   Sequential Subsystem Linkages

In an integrated system these three subsystems would be linked
as follows:

System _ _ _ _ ➤ Subsystem _ _ _ _ ➤ Subsystem _ _ _ _ ➤ Subsystem _ _ _ _ ➤ System
Inputs   1   2   3   Outputs

A variation on this type of nesting of subsystems is discussed
later in the chapter to demonstrate how such flow patterns can be
integrated. Although sequential integration of subsystem flow patterns represents the least complex problem of integration, it is
more common to find that outputs of other systems interface with
certain subsystem transformation processes within a given system
under study. This is analogous to the decision processes within a
given department of a firm which require inputs (the outputs of
other systems) from other departments in the firm.

In addition to the variety of flow patterns which may exist, it is
also possible to have nested feedback responses. A schematic of such
a situation appears in Figure 1-5. The schematic in Figure 1-5 implies that the subsystems have their own feedback loops for correction and control within the subsystem and that there also exists a
higher level of overall feedback for control of the whole system.

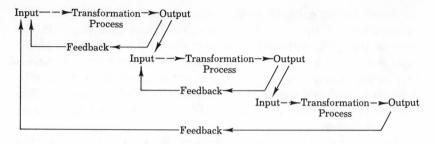

Figure 1-5.   Nested Subsystem Linkages

This is analogous to the feedback reports associated with a department's operation within the firm at the subsystem level and to the higher-level, general management reports which indicate the condition of the firm as a whole.

In the design of a system and its subordinate subsystems it is essential that the logic developed to transform inputs into outputs lead ultimately to the objectives for which the system was designed, without any loose ends or dead ends in the transformation process stream. By block diagramming relationships and identifying inputs, transformation processes, and outputs, this kind of problem can be avoided. Such a procedure, when applied to the existing information systems of firms, usually highlights the illogical branches that some information streams exhibit. Evidence of this is the transformation process which has inadequate inputs or yields outputs which do not lead anywhere and in which no one is interested.

System States

In designing systems it is useful to consider the various states which a system can exhibit. Initially, a system can be said to be in a *static* state. As it is designed, the inputs are identified, the transformation process described, and the outputs identified. However, in the static state, the values of the attributes of the objects do not change; so, in effect, nothing really happens. The concept of static state is useful when one considers the status of the system objects at start time or stop time in its operational cycle.

Between the start time and stop time, when a system is operating and inputs are being transformed into outputs, the system can be said to be in a *dynamic* state. In the dynamic state, feedback becomes quite important because it measures the changes in the attributes of objects over time and provides the necessary correction.

Equilibrium means that the system is operating within the control limits established by the objectives for which the system was designed. *Disequilibrium,* on the other hand, implies that the system is operating beyond the control limits established by the objectives of the system. In the dynamic state a system can exhibit various types of disequilibrium. A typical example of temporary disequilibrium occurs at start-up and shut-down time in many mechanical systems. Although the system starts from its static state and passes through a *transition* state to equilibrium, it falls out of equilibrium through another transition state during shut-down. Such a condition is shown in the abstract terms of a plot of an output object attribute over time in Figure 1-6.

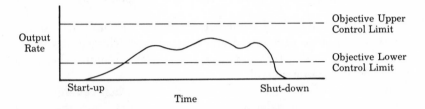

Figure 1-6. Systemic Response to Static–Dynamic State Change

In addition to the problems of start-up and shut-down of the system there are other types of disequilibrium which can develop. One of these is the *exploding* state. In this case the system tends to yield outputs which oscillate and become magnified to the point at which the system has to be shut down. The exploding state is usually caused by the introduction of factors or coefficients which overcompensate to bring about change in a feedback loop. It may also result from delays in the transmission of correction signals. The former problem can often be overcome through simulation of the system during system design before it is put into actual operation. The latter problem usually develops after the system has been in operation some time and is impacted by delays. Figure 1-7 shows the effect of a delay in a correction signal with respect to a system which exhibits a cyclical response pattern of a sinusoidal nature.

In the case of Figure 1-7, the system is being tracked in terms of its output. The signal sent indicates that oscillation is present and that the output is beginning to increase from the trough of the cycle. The presumption is that if correction were immediate then the cycle could be reduced through a damping effect. However, the delay inherent in the transmission of the signal causes it to be received when the actual output is going down near the trough of another

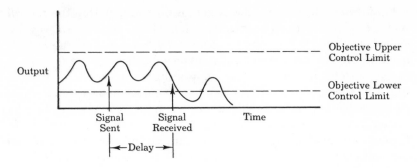

Figure 1-7.   Systemic Response to Signal Delay

cycle rather than up. The correction factor sent indicates that a downward impulse is required and it is applied. This results in the output falling below the objective lower control limit where the system cycles out of control.

An example of this cycle in business occurs when retailers react to a sharp increase in sales of a particular product. They then place orders with their wholesalers who deplete their stocks and in turn place orders with manufacturers. The manufacturers have a manufacturing lead time to produce the product and may have to wait to get supplies from suppliers before they can manufacture the product and ship it to wholesalers who in turn will deliver it to retailers. In this case there are a number of delays and overcompensations working together. The retailers, wholesalers, and manufacturer may add "fudge" factors to their orders which in combination drastically increase the number of products which will eventually be placed on the retailers' shelves. Each of the parties involved is subject to delays in placing orders and carrying out his own functions.

The net result of this is often that the flood of products which were in high demand reach the retailer months later when the demand has all but disappeared. This type of disequilibrium can be overcome by eliminating the overcompensation and sending a single signal all the way from the retailer back to the supplier with the manufacturer and wholesalers informed as well. It is true that not all delay can be eliminated. Physical shipping times and manufacturing process time are cases in point. In many business systems however, the larger delay is not physical but informational. Information delay can be reduced using integrated, data-teleprocessing techniques.

The concept of *damping* is inherent in overcoming fluctuations such as those described above. A system design which incorporates damping in the feedback loop yields just the opposite of the exploding state. When oscillations occur, they are corrected through feedback. Such correction cannot be made immediately without a shut-down of the system; however, it can be corrected over time as the magnitude of the oscillation range is reduced over a number of cycles. Damping is particularly useful in models incorporated in systems which are subject to impacts from the environmental set. A schematic representing systemic damping appears in Figure 1-8.

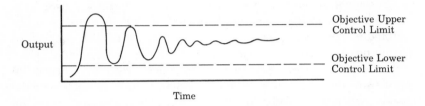

Figure 1-8. Systemic Response to Damping

Another useful concept is associated with the capacity of the system to be *adaptive*. At the initial design stage, a system may be tested for its ability to accept impacts from the environmental set or other systems and return to equilibrium. Such a system is quite useful in that it insures stable attributes of the outputs even though the process or inputs are subject to shocks external to the system. An even more effective system design, however, involves the ability of the system to adapt to changes in the environmental set or other systems and yield outputs which serve the objectives for which the system was designed. In the latter case the emphasis is not maintenance of the original design of the system but rather the automatic modification of system design given external changes which may cause the original system design to be less than optimal in terms of the objectives it serves.

Summary of the Nature of Systems

In developing a basic vocabulary in the systems area one must first consider the definition of a system. Several such definitions are included in this chapter. They have in common, objects, attributes, and relationships. In a general sense a system is a set of objects together with relationships between the objects and between their attributes.

The chapter also includes an examination of the concepts of synergy and symbiosis. Synergy involves the phenomenon of systems outputs yielding a greater value than the sum of the systems inputs. Symbiosis involves mutually beneficial relationships among two or more objects in the system. Implied in this is that the objects involved, when properly related, will yield benefits which would not accrue if such relationships were not reinforced in the systems design.

A distinction is made between systems theory and systems and procedures which suggests that the two are quite different yet, over time, as systems are designed and implemented, the two concepts will undoubtedly reinforce each other. Associated with the process of system design is consideration of the objectives of the system.

The concept of flow patterns is introduced with the conceptual input-transformation process-output scheme characteristic of most system schematics. Subsystems are examined and the potential design requirements for the linkage of subsystems are considered. Inherent in this is the requirement for feedback information.

Finally, system states and system dynamics are considered. Distinctions are made among dynamic states and static states, between equilibrium and disequilibrium of the system, and among unique properties of system states over time such as transition, oscillation, and explosion. In the same vein other dynamic considerations are presented which deal with the relationship of systemic response to signal delays, the concept of damping, and the capacity of a system to be adaptive.

In the next chapter, we shall consider systems analysis, particularly as it relates to managerial decision making. Some consideration will be given to models which can be incorporated in a decision system. In addition, the process of simulation, wherein decision models can be tested for effectiveness both in terms of system design and system analysis, will be discussed. Finally, a procedure for the development of systems is proposed and demonstrated through its application to the development of a simplified inventory management problem.

Conceptual Problems

1. Synergy in business systems is essential to the survival and growth of the firm. It is usually measured as the difference between the

value of inputs (costs) and the value of outputs (revenues) and reported as profits. Many economists have proposed that the objective of businessmen is profit maximization. Many executives would counter with the proposition that profit maximization is neither a desirable nor attainable objective but that reasonable or satisfactory profits provide a more realistic target. Comment on the reasoning behind both of these propositions and support that position which you believe is consistent with the systems concept.

2. In recent years those who have taken a behavioral approach to the analysis of the operation of business firms have suggested that the firm is a social institution which serves the social and psychological needs of its members and that the objectives of the firm should be stated relative to these needs. In the view of some behavioralists, these objectives override the traditional economic objectives commonly encountered in traditional management literature. What is your reaction to this alternative? Does the systems concept provide a vehicle for the integration of both types of objectives?

3. Assume that in an analysis of several subsystem information systems and their linkages you find that certain information outputs (reports) are being produced but are simply filed and never used by management. They reflect a few small white-collar empires which have been developed over the years. How would you go about modifying or eliminating these information outputs and the processes which produce them? Would the necessary actions also involve dissolving the empires? If so, how would you proceed to do this?

4. Some theorists have differed in their views of the desirability of system equilibrium versus system disequilibrium. One group proposes that effective systems activities should yield a homeostatic capability such that the system will revert to a state of equilibrium after being impacted by external or internal forces. The other group proposes that maintenance of the equilibrium state and a design commitment to homeostasis will render the system non-adaptive to important internal and external impacts. In fact, they propose that a certain degree of disequilibrium is beneficial for the firm and that a commitment to a state of equilibrium is tantamount to inviting the development of industrial arteriosclerosis. Examine both positions critically in terms of their ramifications and support the position you believe to be the most beneficial for the firm.

5. Signal delay can be harmful to the effective operation of systems. In inventory systems characterized by oscillating demand patterns, such delays can result in a corrective reduction at a point in time when inventories actually should be increased. What techniques and hardware could be utilized to reduce signal delay from the customer back through the channel of distribution to those responsible for the

allocation of resources committed to the production of particular products?

6. Develop techniques you believe would create a damping influence on oscillating demand patterns in inventory stock levels. How would you predict the number of cycles required to damp an oscillation pattern of $\pm$ 40 per cent to $\pm$ 5 per cent? Would your technique affect frequency of oscillation as well as amplitudes? If so, how would you predict changes in frequency?

Systems Analysis

The Systems Concept and Decision Making[1]

Decision making is central to the process of management, it entails much more than simply the selection of one course of action from many alternatives. Decision making involves a number of other factors when viewed systemically.

To begin with, in order to practice decision making it is necessary to consider the framework within which the decision will be made. The most important element in the framework is the objective of the system. In business this can be interpreted as the objectives of the firm at an all-inclusive level or simply as the objectives in one subsystem of the firm. In either case, the decisions to be rendered must be considered with respect to particular stated desired outputs. Another framework consideration is the set of boundaries affecting the implementation of alternative courses of action. Any firm has certain boundaries or limitations which preclude the development of certain courses of action. These may take the form of limitations in money, materials, people, geographic area, and the like. A third element in the framework concerns value systems. Each decision is rendered with respect to particular values which the people in the firm hold. Generally the predominant value system is an economic one. Managers are con-

---

[1]Adapted from R. J. Hopeman, *Production: Concepts, Analysis, Control* (Columbus, Ohio: Charles E. Merrill Publishing Company, 1965), pp. 64-76.

cerned with dollar inputs and outputs and their related costs and profits. Many other value systems can be considered, however, including certain social values and human values. It is in terms of the mix of these value systems that ultimate decisions are weighed. Finally, the decision maker is concerned with projections of the outputs of his decisions. To arrive at projections of these outputs he may have to resort to the use of deterministic or probabilistic models. If the problem contains elements which are non-quantifiable or if the objectives of the firm are non-quantifiable, or even if the value systems are non-quantifiable, he may have to use judgment and intuition to arrive at his decision. All of these considerations—objectives, boundaries, value systems, and expected outcomes—form the framework within which his decisions are made.[2]

Once the framework considerations have been determined, the decision maker can take a procedural approach. A common one utilizes the scientific method. This method has been used for decades by scientists and includes the following steps:

1. Recognition of a problem
2. Development of a hypothesis
3. Gathering data
4. Testing the hypothesis through experimentation
5. Reaching conclusions about the hypothesis

In the sciences, the problem recognition phase is followed by the development of a tentative statement of what is expected to be the truth concerning the phenomenon in question. This statement becomes the hypothesis which provides a sense of direction for the research effort and gives some indication of the potential research design. This step is followed by the gathering of data.

The scientist then proceeds to set up experiments in which he can test the variables associated with the problem using quantification

---

[2]For further information on alternative approaches to decision making and decision theory see the following: S. H. Archer, "The Structure of Management Decision Theory," *Academy of Management Journal*, December, 1964; K. J. Arrow, "Decision Theory and Operations Research," *Operations Research*, December, 1957; C. P. Bonini, *Simulation of Information and Decision Systems in the Firm* (Englewood Cliffs, N.J.: Prentice-Hall, Inc., 1963); I. D. Bross, *Design for Decision* (New York: The Macmillan Company, 1957); H. Chernoff and L. E. Moses, *Elementary Decision Theory* (New York: John Wiley & Sons, Inc., 1959); W. Edwards, "The Theory of Decision Making," *Psychological Bulletin*, July, 1954; P. Fishburn, *Decision and Value Theory* (New York: John Wiley & Sons, Inc., 1964); M. L. Hurni, "Decision Making in the Age of Automation," *Harvard Business Review*, September-October, 1955; R. C. Jeffrey, *The Logic of Decision* (New York: McGraw-Hill Book Company, 1965).

to measure results. In his laboratory analysis he usually can hold everything constant and test one variable at a time. He utilizes elaborate equipment to monitor physical changes and accurately calibrated devices to measure minute degrees of change.

When the experimentation is completed, a review of the notes may indicate that the original hypothesis was true. If so, this represents not only the solution to the problem but also a statement of expected output should future experimenters or problem solvers wish to check it. In essence, the scientific method in science yields single answers which hold over time.

Although such a method is useful in science, it is questionable whether it can be applied in this form in business decision making. There are a number of differences which require a modified approach. First, problems may not be easy to recognize. Symptoms certainly exist in abundance but getting from the symptoms of problems to the problems themselves is a difficult process. Further, attacking symptoms as if they were problems yields only short-range solutions. The curing of a symptom, in a business sense, is the equivalent of fighting brush fires — the manager quickly dispenses with the symptom of a problem one day only to find that it recurs the next week.

A second area of difference concerns the formulation of hypotheses. In science, the search is for the truth concerning particular relationships. In business, there may be no such truth. Relationships are subject to daily change so that last week's decision may not yield the same results if applied this week. To make decisions effectively in such a setting requires the ability to visualize the relationships of all the pertinent variables as they operate simultaneously. This brings us to the third major difference between business and science. The scientist can hold everything constant and test one variable at a time. The businessman never can. Both the variables internal to the firm and the environmental variables are constantly in flux. Because of this, a useful systems approach must provide a statement of the environmental set, the major system wherein the decision is being made, the subsystems involved, and the inputs, transformation processes, and outputs which are occurring at these interfaces.

A fourth major difference between problem solving in science and in business concerns quantification of data and measuring devices. The sciences place a great deal of emphasis on both these items. But in business, quantitative models may represent only part of the set of relationships involved at a decision point. Other factors which are non-quantifiable may also affect the decision and be quite important. The problem then becomes one of weighing the quantifiable and non-quantifiable factors at the same time. With respect to measuring devices, business, only in recent years, has concerned itself with real-time monitoring of results. Recent developments in information-gathering hardware for business, however, may change this.

A final difference between science and business pertains to the "single answer" concept of science. This is not characteristic of business decision making. There may be several good answers or decisions with respect to the same problem. Indeed, equally competent managers may arrive at substantially different decisions. Again, this results from the complexity of the business environment and the non-quantifiable nature of so many elements in that environment and within the firm.

Even with these differences, the scientific method does have value as a general procedure for business decision making if it is modified somewhat. The following procedure represents such a modification:

1. Recognition of a problem
2. Separation of symptoms from problems
3. Gathering of pertinent information
4. Analysis of pertinent information
5. Developing alternative courses of action
6. Choosing one course of action
7. Follow-up on the decision

As indicated earlier, it is essential for the manager to separate the symptoms of problems from the root problems involved. If he fails to do this, he will get only partial and temporary solutions since he will not have attacked the problem at the root level. As he gathers information he should sift it for pertinence. Today it is possible to gather so much data in a firm that a business manager could never read it all in time to make a decision. Therefore he should provide for computer analysis of much of the data so that he receives only summary reports and exception reports.

After the data have been gathered the decision maker can turn to analysis. In this phase he may use the arsenal of new techniques which have been developed in operations research or many of the more common, established, analytical models in business. He may attempt simulation to develop what is close to an optimal answer. In fact, simulation may be quite useful for the testing of alternative courses of action as different variables are changed. However, in the last analysis, he must overlay the quantitative analysis with judgment, experience, and intuition to incorporate the non-quantifiable variables and to judge alternative courses of action with reference to non-quantifiable value systems and non-quantifiable objectives.

Finally, the decision maker must choose one course of action from the alternatives developed and implement it. Following implementation he should follow-up on the decision to see whether the results being achieved conform to the original expectations. If they do not, then another decision will be required to meet the objectives. This phase, in the systems sense, is the equivalent of developing a feedback loop.

The various approaches which are used in business to make decisions offer useful contrasts for the development of understanding the systems approach and its potential effectiveness. The oldest and perhaps most widely used technique is trial and error. With this technique, the manager identifies the problem, analyzes the situation, and decides on a course of action. He may or may not go to the trouble of developing alternative courses of action prior to his decision. He then monitors the results of the implementation of his decision and, if it worked well, he will use the same decision in the future. If it did not work well, he will reassess the situation and implement another decision. This approach provides quick but not very effective decisions.

The hazards in the trial-and-error approach are many. For one thing, if decisions seem to work well, they may be codified in some sort of set of standard operating procedures or more broadly stated policies. Although this may free the manager from assessing the situation should it arise again and may allow for the delegation of authority for decision making, the unfortunate aspect is that the situation keeps changing while the decisions remain structured and codified. It has been pointed out above that the environmental influences on a decision and the set of environmental constraints may be accurately measured at one point in time when a trial-and-error decision is made. When the same decision is implemented at a later point in time the environmental influences and environmental constraints will have changed. The same problem exists with respect to changing relationships within the firm.

In addition, the trial-and-error approach may preclude the testing of enough alternative decisions to yield "optimal" results. In practice, if a trial decision works, it may be implemented over and over again without ever testing other decisions to see if they will yield better results. Some indications of this hazard are exhibited by the comments of such managers that their approach has worked for years and any change may require taking risks. In short, "don't rock the boat." A similar comment addressed to others who may suggest an alternative approach is something like "we tried that ten years ago and it didn't work."

The systems analysis approach differs significantly from the trial-and-error approach in that the environmental influences and environmental constraints are identified and measured in terms of their impacts at the decision point. In addition, the internal systemic relationships within the firm are identified. At the decision point, all of the significant relationships converge and the relative impacts, both external and internal, are weighed. Instead of adopting the first workable solution, in systems analysis where simulation is used a number of decision rules are tested and the results measured before the best decision is selected and implemented.

Using the systems approach, the problem of rocking the boat does not come up unless the system has been too tightly structured — an ever-present hazard with respect to computerized and programmed decisions. A dynamic, systemic approach constantly rocks the boat by testing alternative values of the interrelated variables until the best solution is found within the number of decision rules tested. Further, the argument that a particular decision did not work ten years ago is irrelevant. The systems concept takes as given that the values associated with particular variables are constantly changing. Thus the decision not to implement a particular course of action is made on the basis that another decision yields better results at the time it is tested. Should a decision be required in the future in this same area, most of the same decision rules may be tested again on the assumption that changing conditions may reveal that a poor decision under past conditions is the best decision under current conditions.

Another approach to decision making, as common as that of trial and error, is that of following the decision patterns of decision makers in other firms or following general principles whether their origins be academic or trade practices. In using this approach the decision maker may wait to follow the lead of decision makers in competitive firms who are faced with similar problems; or he may base his decisions on accepted trade practices. Both approaches amount to "follow the leader."

It has been noted in support of these approaches that the following of principles or the decisions of others tends to minimize risk. It is assumed that if the competitor tries out the decision first, he carries the risk of failure. What is overlooked in this approach is that if the competitor's decisions are effective he will be the first to gain the benefits. It has been said that to become an effective leader, one must first be an effective follower. However, if one remains a follower, he will never be a leader. The manager who bases his decisions on a follow-the-leader pattern will never be a leader or innovator and therefore will forfeit the opportunities available to those who took the bold first steps involved in innovative decision making.

As for following trade practices or principles, the hazard becomes one of accepting sort of an "arithmetic mean" of experience which typically embodies only a fraction of the innovative quality of the original leaders who developed them. This approach often results in what amounts to "backing into the future."

The systems concept overcomes several of these deficiencies. It implies an understanding of the relationships of the variables and constants involved and how these relationships will change in the future. The development of a systemic construct within a firm requires a good deal of custom tailoring for the firm in question if it is to be effective. In custom-tailored system design, the firm, in effect, sets its own patterns of relationships which are not simply copied from others. When the system is in operation, the question is not what shall our *reaction* be to the moves of other firms but rather what shall our *acts* be to alter our relationships with other firms. In a sense, with the systems approach the firm is equipped to act as well as react. When the firm is in a position where it must react, its reaction pattern has been developed in advance to take advantage of faulty or effective decisions of other firms. In this manner it can compete more effectively and aggressively as well as be a leader in its industry.

The final traditional approach presented as a contrast to the systems approach is that in which the decision maker relies primarily on his judgment, intuition, and experience. Whereas the trial-and-error approach implies a sort of tentative testing until something works and the follow-the-leader approach implies non-innovative decision processes, the use of judgment, intuition, and experience is expounded by those who have arrived at successful and somewhat innovative decisions. The approach is also used by those who make miserable decisions; however, they seldom account for this by reflecting on their personal judgment, intuition, and business experience.

This approach is quite effective in many cases but invariably its effectiveness depends on the particular man who has good judgment, uncanny intuition, and vast business experience. If a firm is fortunate enough to have in its management team a number of such men, it is likely that effective decisions will be made which are innovative and bold, yet sound. The only problem with this approach is that it belongs to the man involved, not to the firm, and is only partially transferable. The training of managers with moderate capabilities in these areas, through the use of management development programs, case analysis, business games and so forth, or through the use of understudy programs and job rotation, is only partially effective and very time consuming.

Using the systems concept, the processes of analysis of variables and constants which reflect the state of the system and its interfaces with other objects both within the firm and in the environmental set allow for a more structured, documentable construct. In fact, many effective systems operate well without managerial intervention in terms of the interjection of judgment, intuition, and experience. Those decisions to be made which contain factors which are not quantifiable require subjective overlays. At these points in the system, those people particularly skilled in the utilization of judgment, intuition, and experience play a significant role.

There is no question that with the systems concept we still need managerial skills. The point, however, is that they need not be applied universally in making the decisions of the firm. Many, if not most, of the decisions to be made are routine and programmable in the systems concept. Only a few key, strategic ones require a subjective overlay. Where this is the case, even more time is available for the man with these skills to apply them effectively.

Models

Inherent in the systems concept is the use of models.[3] *Models* represent abstractions of the real world and as such are symbolic

---

[3] Much research has been conducted on models in the management science literature. To facilitate further study of these models a selected listing of useful sources follows: R. L. Ackoff and P. Rivett *A Manager's Guide to Operations Research* (New York: John Wiley & Sons, Inc., 1963); L. E. Arnoff and M. J. Netzorg, "Operations Research — the Basics," *Management Services,* January-February, 1965; L. E. Arnoff, "Operations Research for the Executive," in *Data Processing Yearbook 1964* (Detroit, Mich.: American Data Processing, Inc., 1963); D. Blackwell, *Operations Research for Management* (Baltimore: The Johns Hopkins Press, 1954); E. C. Bursk and J. F. Chapman, eds., *New Decision-Making Tools for Managers; Mathematical Programming as an Aid in the Solving of Business Problems* (Cambridge, Mass.: Harvard University Press, 1963); C. W. Churchman, R. L. Ackoff, and E. L. Arnoff, *Introduction to Operations Research* (New York: John Wiley & Sons, Inc., 1957); J. E. McCloskey, *Operations Research for Management* (Baltimore: The Johns Hopkins Press, 1954).

representations of reality. They typically take the following forms.

Physical models are three-dimensional abstractions of real objects, usually in miniature, which have many of the characteristics of the real objects. Examples of such models include wind tunnels used for testing airframe designs and wave generation troughs used for testing ship hull designs or harbor configurations. In the use of models such as these, a particular airframe design or hull design is subjected to the environmental forces which may be exerted against it in actual use.

The use of such models in systems analysis provides the analyst with a laboratory setting in which he can alter the magnitudes of particular environmental impacts at will. By testing a design against various simulated conditions, he can observe and record results which may provide insights into potential improvements in design. After these improvements are made he can again test the design against identical environmental impacts to get comparable and correllatable readings of the improvements or faults developed in the redesign process. Such models obviously allow for analysis of relationships which could not be undertaken economically with the real-world referents, in this case, aircraft and ships. It is much simpler and cheaper to modify a model airplane or model ship than to modify the real thing to overcome design deficiencies.

Schematic models represent a greater degree of abstraction from real phenomena than physical models. Examples of schematic models are organization charts, flow charts, block diagrams, and other pictorial abstractions. Whereas physical models are constructed to represent objects in a system which are easily observable, schematic models offer greater flexibility in that they can represent relationships which are conceptual. An organization structure does not physically exist, a flow chart does not physically exist, and a block diagram does not physically exist, yet they provide meaningful symbolic representations of relationships which do exist even though they are not physical objects.

Schematic models are particularly useful in systems analysis since such an important part of systems analysis is based on the investigation of relationships which are not physical. They also are of importance in information systems to chart information flow patterns and logic networks in decision processes. Their flexibility is an important asset in building systems since the level of detail desired can be easily modified and integrated both with respect to macro- and micro-systems. An example of this would be a schematic of the environmental set and the relationships of its objects to the firm, a schematic of the information system of the firm with respect to impacts from one object in the environmental set, a flow diagram of the logic for handling such impacts, and finally a detailed block diagram from which a computer program could be written for one section of the flow diagram logic.

Another class of schematic models which is useful in analysis includes graphs and charts of relationships among variables and constants. Such graphs may present, for example, the effects of changes in quantity on cost; the usage patterns of an item over time; as well as other relationships usually depicted on X and Y coordinates. Several examples of this type of model are used in this book.

Mathematical models represent a greater degree of abstraction of real phenomena than schematic models in many cases. Whereas a blueprint or drawing may resemble its empirical referent to some degree, a mathematical model of a particular item looks nothing like it. Such models are totally symbolic representations of reality and as such are often difficult to interpret.

Even though they represent significant levels of abstraction, mathematical models represent very powerful tools of the systems analyst. He is generally interested in relationships among objects and mathematics serves well to define relationships precisely. He is also concerned with the impacts of one or more variables of the system on other system components. Mathematical models serve him well in this area too in that they provide a means for measuring such impacts and analysis of them in common terms, i.e., numbers. For decision purposes, mathematical models provide the necessary flexibility to define a decision rule at one point in time and yet change it at another by altering coefficients or the arrangement of terms.

Within the class of mathematical models are two distinct types which are particularly significant with respect to decision rules incorporated in management information systems. The *deterministic model* is a type of mathematical model in which the variables, constants, and relationships are stated exactly and with certainty. Thus, the utilization of a deterministic model implies that when a given state of variables and constants exists together with definite statements of relationships, the results of the application of the model must be the same. That is, a decision rule which is represented by a deterministic model will yield the same decision in test after test if the relationships, constants, and variables remain the same.

Such models are useful in systems analysis to represent symbolically relationships where conditions of certainty prevail. This, unfortunately, is not the case in most business applications, although it is relatively common in systems in the physical sciences, i.e., chemistry, physics, astronomy, and the applied area of engineering.

The other type of mathematical model employed in systems analysis is the *probabilistic model*. This type of model is used with decision rules where the variables and/or relationships are somewhat uncertain, that is, under conditions of uncertainty. Since many of the decision processes involved in business decision rules involve conditions of uncertainty, probabilistic models play a large role in systems analysis. Unlike deterministic models, probabilistic models do not yield the same answer time after time. Typically one or more variables or set of relationships will be represented by some probability distribution or measure of central tendency and variance. Each time the decision rule is tested, the result will change as the variables and/or relationships which are modeled probabilistically assume different values. In this sense, probabilistic models cannot provide optimal answers in the pure sense as can deterministic models. They do, however, provide quantitative answers with a particular degree of confidence attached, in many cases. What is lost in deterministic precision is usually gained in that the probabilistic model reflects more accurately the real phenomena under investigation. This is characteristic of most models in systems analysis. To get simplification in the model, one usually must give up some degree of accuracy with respect to the real-world phenomenon being modeled.

There are many advantages to using models in systems analysis, but these are often offset by certain limitations. Since a model is an abstraction, it must not be mistaken for the real-world referent in the system. Some analysts become so enamored of their models that they continue to enrich them to the point where they believe the model is the real world. Or, in other words, the analyst may come to trust his model more than its real-world referent and base decisions on symbolic representations rather than on the real interaction patterns which they represent.

Another limitation concerns the ability of the systems analyst to incorporate in a model all of the important variables, constants, and relationships which exist in its real-world referent. In many cases in business, the number of variables, constants and relationships involved with respect to particular phenomena far exceed our ability to incorporate them in equations of manageable proportions. Some variables and relationships may be non-quantifiable and they are therefore totally excluded from the equations in the model. Although this is a definite limitation, if it is recognized, it can be compensated for by using management judgment to provide the necessary subjective overlay. The real hazard is that management may assume that the mathematical model represents the whole set of relationships and variables involved. Seldom are real situations as unambiguous as mathematical models of them imply.

When models are incorporated in systems analysis, their real value lies in the flexibility they provide to analyze complex relationships. The ultimate test of their incorporation in a management information system, particularly within decision rules, rests on how well they represent the real-world referent. To be effective tools, they should provide both better understanding of the dynamics of the real-world phenomenon being modeled and predictions of future system states as real-world variables and relationships are altered. If they provide these things, they become particularly effective in improving decision making.

Simulation

Simulation involves the experimental use of models to study the behavior of a system over time. As such, it can be used to explore alternative system designs or to explore the behavior of the components of a system to better understand their relationships within the system.[4]

One approach to decision making, discussed in a preceding section, involved trial and error. Such an approach is similar to simulation in that alternatives are explored using experimental tests of alternative configurations of the attributes of variables, of variables themselves, or of relationships among variables.

---

[4]For further information on simulation see the following: D. N. Chorafas, *Systems and Simulation* (New York: Academic Press, Inc., 1965); H. S. Krasnow and R. A. Merikallio, "The Past, Present, and Future of General Simulation Languages," *Management Science*, November, 1964; W. H. Lawson, "Computer Simulation in Inventory Management," *Systems and Procedures Journal*, May-June, 1964; D. G. Malcolm, "System Simulation—A Fundamental Tool for Industrial Engineering," *Journal of Industrial Engineering*, May-June, 1958, and "The Use of Simulation in Management Analysis—A Survey and Bibliography," *Operations Research*, Vol. 8, (1960); E. W. Martin, ed., *Top Management Decision Simulation —The A.M.A. Approach* (New York: American Management Association, 1957); J. H. Mize and J. G. Cox, *Essentials of Simulation* (Englewood Cliffs, N.J.: Prentice-Hall, Inc., 1968); T. H. Naylor, J. L. Balinfy, D. S. Burdick, and K. Chu, *Computer Simulation Techniques* (New York: John Wiley & Sons, Inc., 1966); K. D. Tocher, *The Art of Simulation* (Princeton, N.J.: D. Van Nostrand Co., Inc., 1963); *Bibliography on Simulation*, IBM Manual 320-0924-0 (White Plains, N.Y.: International Business Machines Corporation, Technical Publications Department, 1964).

In the trial-and-error approach the variables and relationships of the real-world system are altered, decisions are actually implemented, and feedback of results is obtained from the real-world system. In simulation, the analyst implements decisions in an abstract system which represents the real-world system. Since the commitments made by decisions in simulation are made "on paper" only, they do not involve the risks associated with actual commitments. In essence, the decisions can be tested artificially and results evaluated prior to the actual commitment of resources or alteration of relationships within the system.

Another contrast concerns the manipulation of time. In a trial-and-error approach, actual calendar time is expended to test alternatives. This may be a relatively slow process. Using simulation, the calendar time periods can be artificially compressed by making simulation runs over thousands of time periods in a single day. It is in this sense that simulation provides the opportunity for greatly accelerated dynamic tests of models.

Another approach to decision making previously discussed involved the use of the scientific method applied to business. Whether this method involves subjective analysis of the problem or the use of elaborate deterministic mathematical models in the analysis phase, the condition implied is that the system is in a relatively static state. That is, given a certain set of static conditions, analysis would indicate that a particular course of action should be followed.

The use of simulation differs in this case not with respect to the use of models but rather with respect to the assumption of a static state of the system. In simulation, it is assumed that the models are tested dynamically over time, subject to the exogenous and endogenous variables which impact the system. Further, in many cases these impacts are generated probabilistically rather than deterministically.

Simulation implies that the objects, attributes, relationships, and parameters of the system are described in some fashion, usually quantitative in nature. A deterministic model implies the same things; however, the result expected from a deterministic model is a single answer, usually dubbed by hopeful model builders the "optimal" solution. The use of simulation, on the other hand, does not yield the optimal answer but one which is judged to be best according to the objectives for which the system was designed and within limits of the number of experiments executed with the simulator. The process of simulation is depicted in Figure 2-1.[5]

---

[5]Adapted from Joe H. Mize and J. Grady Cox, *Essentials of Simulation* (Englewood Cliffs, N.J.: Prentice-Hall, Inc., 1968), p. 7.

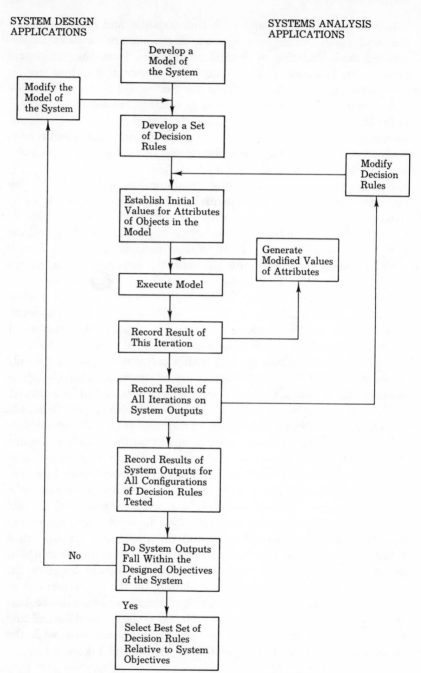

Figure 2-1.   The Process of Simulation

At the initial stage of simulation, the real-world phenomenon under study is abstracted and symbolically represented by a model which states the objects in the system, their attributes, and the relationships among them. The nature of the relationships is then stated in terms of decision rules which indicate what course of action should be taken, given a particular state of the system.

Initially some quantitative values must be established for the attributes of the objects in the model, i.e., initial values assigned to the variables in the equations. In conjunction with the stated parameters and the decision rule under test, these variables are incorporated in the execution of the model or solution of the equation. A record is then made of the results of the mathematical operation for time period n.

The time period is then advanced to $n+1$ and the values associated with the next time period are generated to represent changes in the values of the attributes. The model is again executed and results recorded. This loop continues to operate through the necessary number of iterations until the total number of time periods being simulated has been reached. At this point the result of moving the model through a number of time periods with a given set of decision rules is recorded in terms of system outputs.

The next phase involves modifications of decision rules. After the first set of decision rules has been tested over the number of time periods, with changing attribute values, a different set of decision rules is tested in the same manner. This outer loop is then repeated for all decision rules to be tested. The simulation process thus involves an inner loop of changing attribute values over time within an outer loop of changing decision rules. The outer loop is analogous to testing alternative courses of action in the scientific method as applied to business decision making. The operation of both loops represents systems analysis applications of simulation.

After the outer loop has been executed and all configurations of decision rules have been tested, the results are examined, in terms of system outputs, to see if they fall within the objectives of the system. If they do, then that set of decision rules which yields outputs which most closely meet the objectives of the system is selected and implemented. If they do not, then the model of the system may need to be redesigned. In such a redesign phase, modifications may be made in the objects, attributes, and/or relationships represented by the model.

Considering the schematic of the process of simulation it can be seen that simulation serves two purposes: systems analysis, and improvement of systems design. As decision rules are changed and simulations are run to test them, the systems analyst is provided with information about the behavior of the system. This information is quite useful in the analysis of how and why the system operates. It gives some indication of the importance of particular variables and relationships in meeting system objectives.

In the broader sense of systems design, simulation provides insights into potential areas for improvement of the models of the system. If a current design does not yield appropriate outputs, then additional objects may be added to the system, different attributes recorded, or alternative relationships established. From simulated experiences with a given system, results may indicate that particular variables are more sensitive than others in generating outputs and, thus, these may be accentuated in a new design while others are removed from consideration. Testing a system design through simulation may also indicate that particular impacts of the system from the environmental set have been ignored and, in redesign of the system, they should be incorporated.

A Procedure for the Development of Systems

Business systems are invariably complex. Indeed, their complexity is so great that even though it would be challenging to design a total system for the entire firm which would be operationally effective, this is not really possible, given our level of understanding of the dynamics of business management. If it is unlikely that total systems can be developed at the present state of the art, then what should be the design objective of the systems analyst?

A workable design objective is the design of a subsystem of the firm. In situations where much is known about *what* exists and *why*; where the variables and constants involved in the phenomenon can be isolated; and where their impacts can be measured by sensors; then there is a potential for effective system design, even though the design is on a relatively restricted phenomenon and may need to be adjusted later in time as it is linked to other subsystems, yet to be developed.

There are many ways to design a system. They have in common the development of nested input-transformation process-output constructs at the subsystem level. These subsystems interface with other systems and the environment. They all appear to be logical and orderly and have characteristic flow patterns. Yet, how a person or team develops such a set of relationships has not been distilled into a standard methodology. Although several methodologies are discussed in the literature, no one approach can be said to be the best. The methodology discussed below is similar to others, yet it may be useful to the reader who has not developed a systems methodology of his own.

1. Evaluate the problem and state objectives.
2. Define the environmental set.
3. Define the system encompassing the basic transformation processes associated with the problem.
4. Define the subsystems.

System Conceptualization

5. Apply appropriate techniques to optimize the subsystems.
6. Test subsystem models.
7. Link subsystems to form the system.
8. Apply appropriate techniques to optimize the system.
9. Test the system model.
10. Modify the model for environmental influences.

Model Construction and Simulation

11. Document logic and state parameters within which the dynamic system is reliable and valid.

Documentation

The first phase of systems design involves a clear statement of the problem or problems to be solved. This involves isolation of the variables and constants associated with it and the parameters within which a solution would be feasible. The design should include a statement of the objectives to be achieved — What exactly is the system supposed to accomplish and in what manner will its success be evaluated? If possible these objectives should be measurable (i.e., quantitative rather than qualitative) so that alternative system designs and system performance can be measured for effectiveness.

The next phase of a systems study involves system conceptualization. With knowledge of the nature of the problem and the proposed objectives of the system, a useful first step is to construct the relationships of the objects of the system together with their attributes in a given state, at a macro or environmental level. One approach to this involves determining all of the environmental factors which have a direct influence on the system in terms of inputs or which are affected by the outputs of the system.

Once the environmental set is conceptualized, it is useful to define the system. Since a system, its environment, and subsystems are classifications based on a particular vantage point, it can be difficult to define the particular system which involves the problem in question. In general, the system can be viewed as that construct which encompasses the basic transformation processes to transform the most important inputs into the most important outputs.

As the system is enclosed by its environment, so the subsystems can be viewed as subsets of the system. Once the construct of the system has been developed, it is useful to define the subsystems of which it is composed. In most cases the conceptualization involved in developing these three levels of systemic constructs — environmental, systemic, and subsystemic — will take the analyst from large, complex, ill-defined relationships to the point where his attention can be centered effectively on relationships which are reasonably well defined, less complex, and smaller in proportion, i.e., manageable in an operative sense.

Once the system conceptualization phase is completed, the analyst is in a position to begin model construction and simulation to test the effectiveness of the models. With the subsystems defined, he is able to examine each subsystem in terms of the inputs, transformation processes, and outputs involved. Since each subsystem is closely defined and involves a relatively limited number of inputs and outputs, the analyst is in a position to bring suitable analytical techniques to bear on its effective operation at critical decision points. Once he has constructed a model for the subsystem, he can test it, utilizing simulation, to see whether it yields the desired results under varying conditions. This dynamic test may result in information which indicates that the model requires modification to achieve the desired results. After several iterations, an effective subsystem model should result, and any limitations relating to its parameters, reliability, or validity should then be noted.

While models are being constructed and tested through simulation for each of the subsystems, the definition of a particular subsystem may be called into question. If this happens, it may be necessary to modify the conceptualization of particular subsystems and start model construction over again.

The next major task facing the systems analyst involves the linking of subsystems with respect to the system under analysis. It is in this phase that inputs and outputs of subsystems are carefully correlated to see that they are consistent, complementary rather than conflicting, and reinforcing. As the subsystems are linked together, relationships are being established which should bear a close resemblance to the conceptualization of the system. In many cases, these relationships may involve the incorporation of appropriate techniques to optimize the operation of the system, in addition to the simple linking of the outputs of one subsystem to the inputs of a successive subsystem.

When the subsystems have been linked in a model, it can be tested to see whether it yields appropriate system outputs. If not, the relationships among subsystems may need modification, the linking techniques may need alteration, or the system may require redefinition at the conceptual level. As indicated before, once an effective systems model has been tested through simulation, the logic of the system, its parameters, reliability, and validity should be noted.

Once the systems model has been constructed and tested, it is possible that environmental influences will exceed its design capacity in terms of inputs and/or outputs. Again, simulation can be used to measure the model's capacity for varying inputs and their magnitudes. It also can be used to determine whether the outputs are variable enough to meet changing environmental requirements. If the model does not achieve this, a modification of the model may be required or the definition of the environmental set, at a conceptual level, may be in order.

Whereas environmental influences were difficult to perceive at the outset of the systems study, the analyst now knows where to focus his attention in terms of the operation of the system. He has developed a detailed frame of reference where environmental impacts can be traced to follow their ramifications. He may wish to introduce various influences from the environmental set on the system under study and determine the magnitudes and effects of their impacts. Such experimentation may yield valuable information as to further modification of the system so that it can capably handle a variety of these influences.

As the final step in the procedure for the development of systems, the analyst should document the logic used in establishing the system. This is best achieved through the use of block diagrams. These clearly indicate the inputs, processes, and outputs of the system and the subsystems. They also provide a schematic of the nature of the integration of the subsystems. In addition, it is important that the analyst note the parameters within which the system is designed so that, when conditions beyond the parameters exist, he is aware that the output of the system does not represent a proper evaluation of them.

An Example of the Use of the Systems Development Procedure

As an example of the process of systems conceptualization, model building, and simulation, consider a greatly simplified inventory management problem. Deeper and more elaborate analysis of this problem appears in Part II of the book, but at this point a broad-brush treatment will illustrate the general procedure.

The problem of inventory management will be stated as providing raw material inventories of one thousand different items sufficient to keep production facilities in operation without shut-downs and yet minimize costs. The objective can be measured in terms of cost and the whole management program is intended to be automated in terms of the information system. In a more complete inventory system, the concept of materials flow from suppliers through the firm to customers would be considered, but for purposes of illustration this example is limited to the supplier-production interface within the firm.

An examination of the business environment might include such groups as banks, stockholders, government, customers, competitors, the community, labor unions, and suppliers. For purposes of simplicity in this example the key environmental component is the group of suppliers and their relationship with the firm. Other groups which have significance include banks, which may need to supply funds for carrying inventory; government, which may, through regulation, affect priorities for materials, modes of shipment, tax motivations and the like; customers, whose demand patterns may affect the amount of inventory in the firm, and the balance of these inventories as they reflect changing demand patterns of finished products; competitors, as they may alter service policies to accelerate throughput of products; and labor unions as they may affect plant operations of the firm, its suppliers, and the transportation facilities used to move raw materials inventories from the suppliers to the firm.

Needless to say, the environmental set, although identified in terms of the groups above, is characterized by sets of relationships which are very difficult to define. Even more perplexing is the problem of control. One is aware, for example, that a strike by drivers of trucks delivering raw materials will drastically affect inventory management decisions, but the probability of this event, and dozens of others like it, is difficult to formulate and relate. Thus, for the moment, it is sufficient to recognize the objects of the environmental set without giving a great deal of thought to their relationships or the values of particular attributes associated with these objects.

The system involved in this hypothetical problem pertains to the function of inventory management. In essence it involves groups within the firm such as purchasing, accounting, production, marketing, shipping and receiving, storeskeeping, and so forth. In a larger systems study a materials flow concept might modify such a traditionally functional organization structure; however, for this example, traditional functionalism will be retained as a familiar frame of reference.

Each of these groups is related to one or more of the others in a manner which provides the firm with an information system pertaining to inventory management. For example, marketing may provide information pertaining to consumer demand which may work back through the production function and emerge as production schedules. The production schedules determine what is to be produced and when it will be produced. This information, in turn, indicates the raw materials inventory requirements. Storeskeeping is responsible for maintaining the raw materials which are received by the shipping and receiving department. Purchasing is the liaison between the suppliers and the firm and requires information of the raw materials to be supplied, their quantities, and their due dates. Accounting collects, analyzes, and distributes cost information relative to raw materials inventory.

Without going into the detail of the systemic construct involving these areas of the firm, it should be apparent that the information requirements, in terms of inputs, transformation processes, and outputs, could be established among them. Fundamentally, one must determine in each case what the information inputs are, how they are transformed into information outputs, and where they go for decision purposes among other areas of the firm.

Let us now turn our attention to the subsystem conceptualization stage. Only a few of these will be considered in this example. Of the many questions associated with inventory management, three are common and have an important bearing on the operation of the system. First, how much of each item is in inventory at a given point in time? Second, when should the order be placed? Third, how much should be ordered from suppliers? We shall examine each of these questions as if it were a subsystem of the inventory management system. In each case the inputs, transformation processes, and outputs will be determined and block diagrammed.

Stock Level Determination

In a simplified example we can consider that the *inputs* to a stock level determination subsystem include the following: current level of inventory of each item in units at the beginning of the time period, quantity of each item received over the time period, and quantity of each item used over the time period. Each item could be designated by a stock number.

The *transformation process* involving determination of new stock levels at the end of a given time period would amount to adding the units received during that period to the stock on hand at the beginning of the period and subtracting the units used.

The *output* would be the quantity of each item in inventory at the end of the period. One might also want to consider units on order and in transit plus certain allowances for returns to stock, shrinkage, damaged material, and so forth. But to keep things simple, we shall overlook these matters for the moment.

A schematic diagram of the information flow through the subsystem appears in Figure 2-2.

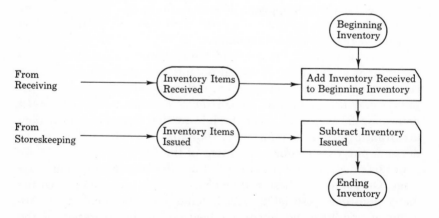

Figure 2-2.   Stock Level Determination Subsystem

Reorder Point Determination

In the subsystem concerning reorder point determination, an answer is sought to the question, When should an order be placed with a supplier so that the production facility does not run out of stock in the interim? One approach to this problem is to compute reorder points. The technique which is applicable requires the following *inputs*: the quantity which is normally ordered; the usage rate per day at which this quantity is consumed; the lead time required to place the order, get it filled, shipped, and received at the plant; and a safety stock level which insures against running out of stock should an order be delayed or demand increased. The basic construct of this model is shown in Figure 2-3.

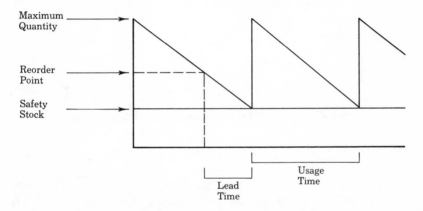

Figure 2-3. Reorder Point Determination Model

Given the quantity to be ordered, the *transformation process* involved is to quantitatively determine the reorder point. To do this, one must first determine the usage rate. This can be found by subtracting the safety stock from the maximum quantity which yields the order quantity. Then the order quantity is divided by the usage period which yields the usage rate in units per day. The usage rate is then multiplied by the lead time and the product is added to the safety stock. The resulting *output* is the reorder point in units which indicates that if an order is placed when the inventory reaches that level, it should arrive just when the raw materials inventory reaches the level of the safety stock.

This simple model assumes linear usage rates, a uniform safety stock, constant lead times, and instantaneous replenishment. Any or all of these assumptions may be incorrect in a particular application and, if so, would require modification of the analytical technique to improve the reliability and validity of the model. In this simple example, however, we shall assume the above assumptions hold true.

A schematic diagram of the information flow through this subsystem appears in Figure 2-4.

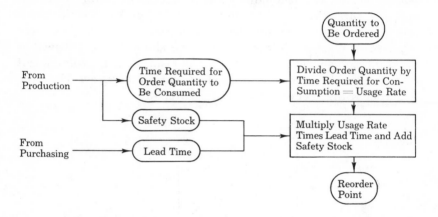

Figure 2-4.   Reorder Point Determination Subsystem

Economic Order Quantity Determination

In determining how much to order, two types of incremental costs must be considered. The first is the procurement cost, which includes costs of making requisitions, analysis and selection of vendors, writing purchase orders, following up on the orders, receiving materials, inspecting them, storing them, updating inventory records, and carrying out the necessary paperwork to complete the purchase transactions.

These costs can be accumulated and can be treated as a fixed cost per order. Since procurement costs per order are assumed to be fixed, the procurement cost is reduced, on a per unit basis, when large orders are placed. Thus to minimize this cost, the largest possible order should be placed. It would equal the total requirements. This practice, however, overlooks the second important incremental cost, carrying cost.

The inventory carrying cost includes a number of items: interest, taxes, obsolescence, deterioration, shrinkage, insurance, storage, handling, and depreciation. The larger the average inventory on hand, the larger these carrying costs become. So in order to minimize carrying costs, one would tend to carry very small inventories. It can be seen that the managerial objectives pertaining to procurement costs and carrying costs are conflicting and therefore that some compromise must be reached. That compromise involves consideration of total incremental cost for both factors and purchase of that quantity which represents the minimum total incremental cost. This situation is depicted in Figure 2-5.

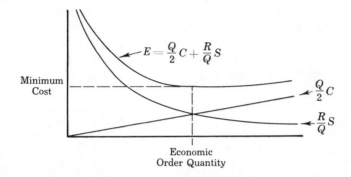

Figure 2-5. Incremental Cost Relationships

In the determination of the economic order quantity, no mention has been made of the effects of quantity discounts, changes in prices, or changing values of any of the cost factors over time. In a sophisticated model, these and other factors would have to be included.

In the simple model used in this example, an equation can be used to determine the economic order quantity to purchase, which will minimize costs.

$$Q = \sqrt{\frac{2RS}{C}}$$

Another equation can be used to determine the cost of purchasing in that quantity.

$$E = \frac{Q}{2}C + \frac{R}{Q}S$$

In effect, the cost equation determines the average inventory $(Q/2)$ and multiplies it times the inventory carrying cost per unit per year (C). To this it adds the procurement cost by multiplying the number of orders required per year (R/Q) times the cost of placing an order (S). Thus for any purchase volume (Q), the cost can be determined and by differentiating the cost equation, one can determine the minima Q through the quantity equation.[6]

In order to use this analytical technique, it is necessary to determine from marketing the annual requirements of finished products and to project these back to raw materials requirements. It is also necessary to gather from accounting, data on the procurement costs per order and carrying costs per unit per year for each raw material item in inventory. Given these *inputs*, the first equation can be used to determine the quantity to purchase and the second equation can be used to determine the incremental cost for each item in inventory. The use of these equations represents the information *transformation process* and the economic order quantities and associated costs represent the *outputs* of the subsystem. A schematic diagram of the information flow through the subsystem is depicted in Figure 2-6.

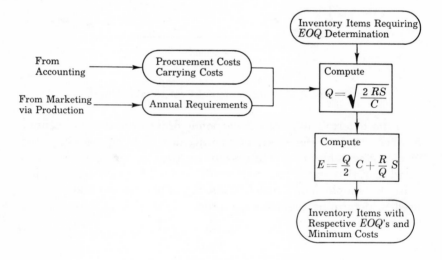

Figure 2-6.   Economic Order Quantity Determination Subsystem

---

[6] The derivation of the quantity equation from the cost equation is as follows:

$$E = \frac{Q}{2} C + \frac{R}{Q} S \qquad \frac{C}{2} - \frac{R}{Q^2} S = 0$$

$$\frac{dE}{dQ} = \frac{C}{2} - \frac{R}{Q^2} S \qquad Q = \sqrt{\frac{2RS}{C}}$$

minimum when $\qquad\qquad \dfrac{d^2E}{dQ^2} > 0$

Linking the Subsystems

The analysis of the preceding three subsystems, relating to stock level determination, economic order quantity determination, and reorder point determination, answers the three basic questions inherent in the inventory management system: how much of each item is in inventory at a given point in time, how much should be ordered from suppliers, and when should the orders be placed.

In a systems study, each of these subsystems could be improved by testing the model with simulated data to determine if it could effectively handle unusual situations. Assuming that the information systems which have been developed in each of these three subsystems operate effectively in terms of inputs, transformation processes, and outputs, the next problem becomes one of subsystem integration or linking to form the inventory management system.

One approach to this would be to design a computer program which, at the end of a given time period, for example a week, would first compute the economic order quantity for every item. This quantity is the output of the EOQ determination subsystem. This output would then be linked to the reorder point determination subsystem, as one of its inputs, to determine the reorder point for each inventory item. The output of that subsystem would then be used in conjunction with the stock level subsystem to determine if the stock level of any item, at the end of the time period, was below the reorder point. If certain items were below the reorder point, these would be recorded with their respective order quantities as outputs of the program. These outputs could then be used as inputs to another system, e.g., purchasing, which could automatically select vendors, place purchase orders, and confirm, for purposes of inventory management, that the order for replenishment of stock had taken place.

Such a linkage as described above would analyze periodically all items in raw materials inventory for EOQ and reorder point. On a daily basis or other short-time interval it would check inventory status against the reorder points and signal purchasing for replenishment when required. Such a daily scan of inventory would operate as a stock monitor and should maintain a state of general equilibrium as a system with only exception reports for replenishment exiting the system to interface with purchasing. A schematic of the simplified inventory management system composed of the linked subsystems is shown in Figure 2-7.

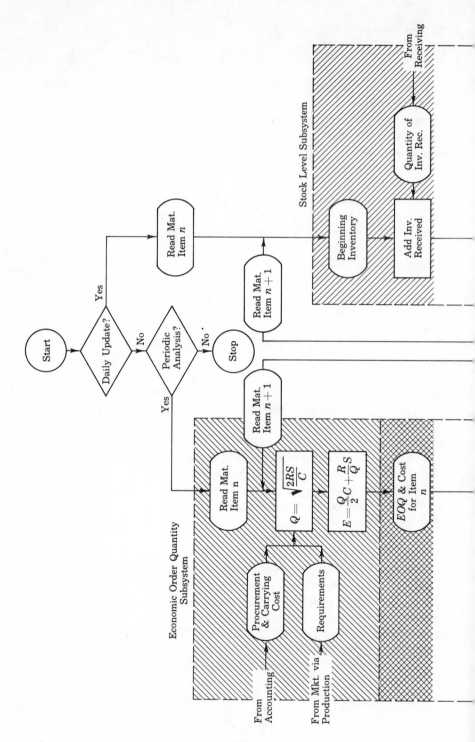

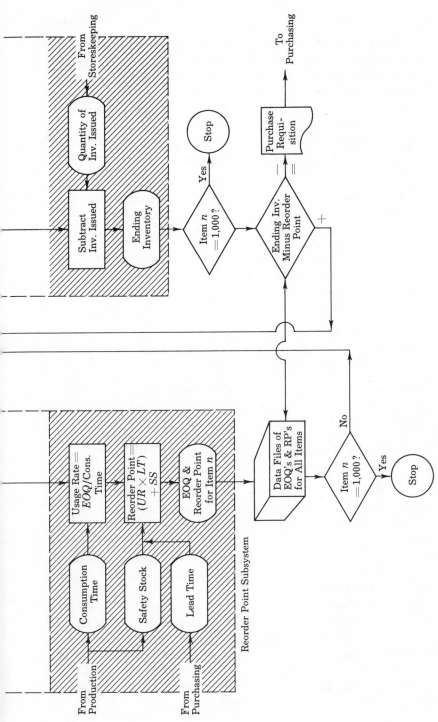

Figure 2-7. Raw Material Inventory System

The test phase of the system model involves simulating decisions by utilizing operating data. In such simulation, it is possible to determine if certain systemic constructs have been overlooked or improperly structured and also to determine if the analytical models have been integrated in such a fashion as to yield optimal answers in an appropriate form for management.

To go through one simulated test let us assume that the inputs of the first subsystem (economic order quantity determination) are as follows:

> Annual requirements = 1000 units
> Inventory carrying costs per unit per year = $0.16
> Procurement costs per order = $20.00

The *outputs* include an economic order quantity of 500 units and an incremental cost of $80.

The second subsystem (reorder point determination) assumes *input* values as follows:

> Usage rate = 10 units per day
> Safety stock = 100 units
> Lead time = 10 days

The *output* yields a reorder point of 200 units.

The third subsystem (stock level determination) operates during daily updates and assumes the *input* values that follow:

> Beginning inventory = 250
> Receipts = 0
> Issued = 50

The *output* yields a current inventory level of 200 units.

The compare function tests the reorder point (200) and the current inventory level (200). If the inventory level is equal to or less than the reorder point, a purchase requisition is printed. In this case the order would be printed for stock number 001 listing the *EOQ* at 500 units.

From this point on, purchasing would place the order for 500 units. When it arrived in 10 days, the stock would be at the safety level and would be replenished in time to avoid stockouts without exceeding maximum inventory levels. Furthermore, the amount ordered would yield the lowest incremental cost. Thus, in this simplified example it would appear that the system functions properly with respect to its objectives.

Note that it has been designed not to produce a great deal of paperwork. Only exceptions are reported. If further information is required by a manager he can retrieve it from the computer's storage unit via the console with relatively simple search programs.

Simulation of the System

The logic involved in these three subsystems begins with the determination of the most economical quantity of the inventory item to order. As an example we shall use the *EOQ* concept with the following values as inputs. Annual requirements are determined to be 40,000 units to be used over a 250-day manufacturing year. The ordering cost per order is determined by accounting and procurement personnel to be $100 for the rather complex ordering procedure. The carrying cost supplied by accounting is 20 per cent of the price, which is $0.40 per unit or $0.08 per unit per year.

Given these cost inputs from various sources, the *EOQ* subsystem yields the outputs, *EOQ* and incremental cost at the *EOQ*, using the following transformation process.

$$EOQ = \sqrt{\frac{2RS}{C}} = 10,000 \text{ units} \qquad E = \frac{Q}{2}C + \frac{R}{Q}S = \$800$$

Although the information system would retain the incremental cost for this stock item, the *EOQ* would be forwarded to the reorder point subsystem to determine when orders should be placed.

In the reorder point subsystem, certain items of input information are required. One of these is the *EOQ*, in this case 10,000 units. Another is a safety stock assumed to be 2,000 units. A third item of information is the usage rate which can be determined by dividing the annual requirements (40,000) by the number of work days (250). This yields a usage rate of 160 units per day. Dividing the *EOQ* by the usage rate would yield a usage period of 62.5 days during which the *EOQ* would be used in the production process.

Thus, four cycles of inventory replenishment would occur during the year. The final information input of the reorder point subsystem is the lead time to place an order, get it filled, and shipped. This will be assumed to be 15 days.

Given the input data above, the reorder point subsystem yields the output reorder point of 4,400 units, according to the following transformation process.

$$\text{Reorder Point} = (UR \times LT) + SS = 4,400$$

The outputs of 4,400 units from the reorder point subsystem and 10,000 units from the economic order quantity subsystem serve as inputs for decision purposes in conjunction with the stock level subsystem.

The stock status subsystem uses as inputs the inventory received during each time period and inventory issued during each time period. It computes the ending and beginning inventories. The subsystem operates in an iterative fashion, updating inventory levels until the reorder point is reached. Then a replenishment order is generated to bring the stock back to its maximum level after the lead time has expired. To indicate how this works we shall trace the stock status conditions as the subsystem operates through time as indicated in Table 2-1.

TABLE 2-1

Time Period	Beginning Inventory	Receipts	Issues	Ending Inventory
0 days	2,000	10,000	0	12,000
10 days	12,000	0	1,600	10,400
20 days	10,400	0	1,600	8,800
30 days	8,800	0	1,600	7,200
40 days	7,200	0	1,600	5,600
47.5 days	5,600	0	1,200	4,400 = Reorder Point
50 days	4,400	0	400	4,000
60 days	4,000	0	1,600	2,400
62.5 days	2,400	0	400	2,000 = Safety Stock
63 days	2,000	10,000	160	11,840 = Stock Replenished

This simplified example indicates that, under the assumptions given, the three subsystems can be linked and the information generated can be maintained internally in a computer. The only output need be signals to place an order. The signal is generated by the stock status subsystem sensing the reorder point which will, in turn, cause an order to be placed for the *EOQ* to minimize costs. This order should arrive by the time the regular stock reaches the safety stock level.

Although the foregoing inventory management system is simple and seemingly applicable to many situations, it is fraught with many limiting assumptions. For example, the economic order quantity subsystem assumes that the annual requirements for any inventory item are known with certainty. This is seldom the case, and thus attention should be directed toward techniques which will provide such a forecast given the characteristics of uncertainty which normally surround this problem.

The EOQ subsystem also assumes that the inventory carrying cost per unit per year is known. Typically, it is not and is subject to gross estimates. Attention should be given to how the determination of such carrying costs can be made so as to make the model more accurate in terms of its relationship to actual operating conditions. In addition, the ordering cost component of the EOQ subsystem is treated as a given dollar amount per order placed. This also is not so simple in actual operations. Further considerations should be explored in this case.

In the case of the EOQ subsystem it is assumed that the optimal answer resulting from the EOQ equation is the order quantity which would be placed under any conditions. In fact, the slope of the incremental cost curve may indicate that considerable latitude exists wherein, in many cases, the EOQ may be adjusted either lower or higher without significant increases in cost.

Further, consideration must also be given to questions concerning quantity discounts, potential price changes for inventory items, obsolescence, multi-shipments, and conditions wherein items are transported to the company in other than single replenishment cycles.

Similar limitations exist with respect to the reorder point determination model. The model discussed at this point assumes linear usage rates. Few industrial situations exhibit this. Usage rates generally tend to vary over time, increasing on some days and decreasing on others. Ways of modifying the reorder point determination model to account for this will be discussed in Chapter 9. However, for the moment, the model presented has a distinct limitation due to its assumption of linear usage rates.

The model also assumes a fixed lead time. In many cases this may not be representative of the phenomenon being modeled. For some items, the lead time may be fixed, but for others it may be variable. The variance can be represented by probability distributions, and expectations of such variance can be simulated; however, this modification is not included in the model presented with the reorder point determination subsystem.

Another limitation concerns the safety stock. In this model it is given as a fixed amount. In an effective model, some care must be given to the determination of the safety stock. The process of determination may involve analysis of past stockout patterns and changes in demand-productive response patterns. Probability analysis plays a definite role in this area and, since this is the case, it may be useful to consider flexible safety-stock levels rather than a static one as is assumed in this model.

Many other limitations could be mentioned concerning the simplified system described here. However, the discussion of some of the limitations described above serves to indicate the kind of considerations to be given the system with respect to the notation of its reliability and validity. This notation is a crucial step in the systems development procedure.

Summary of Systems Analysis

In this chapter the systems concept and decision making is discussed. The scientific method is examined and adapted to fit the requirements of decision making within a business context. Then several current approaches to business decision making are contrasted to the systems approach. These contrasts indicate that the systems approach provides an effective alternative which overcomes many existing deficiencies in current decision processes.

Since models are used to some extent in decision systems, consideration is given to the development of a basic understanding of their nature as well as their characteristic types. These include physical models, schematic models, and mathematical models of both the deterministic and probabilistic varieties.

Simulation, an effective tool in systems analysis and system design, is examined. Its uses are explored and the requisites for its application are considered. A general schematic of the process of simulation is presented which provides a guide for systems analysis and system design. This guide provides a general overview which is of use later in the book when one is immersed in the detail of models and tends to overlook the ultimate tests of their effectiveness through simulation and later in actual application in decision processes.

A procedure for the development of systems is presented which involves three major activities which, in turn, involve several more detailed activities. These three major activities are 1) system conceptualization, 2) model construction and simulation, and 3) documentation. The use of this procedure is then demonstrated through its application to a simplified inventory management system. Each step is carried through in the example to clarify how and why a systemic approach to the development and analysis of systems can be more effective than attempts at aggregate generalization or myopic model building.

The procedure itself parallels the plan of the book. Consideration is given to the broad environmental set, then to the system (the firm), and then to the subsystems (material flow network and its four associated subsystems). This level of consideration is conceptual and proceeds from broad concepts to narrower ones within it until that point is reached where application of concepts and theories can take place meaningfully.

Once at the subsystem level the procedure is reversed and models are developed at the subsystem level, then the firm level, and finally the aggregate firm model is tested against environmental impacts. Testing at each of these levels often involves the use of simulation discussed earlier in this chapter.

In the remainder of Part I of the book we shall be dealing in system conceptualization. At this point the reader has the necessary concepts and vocabulary to work effectively with such broad conceptual schemes as general systems theory at the broadest level, with the environmental systemic construct at a narrower level, and with the firm as a system at an even narrower level. Thus at this point, we take a significant departure from basic system design and systems analysis considerations to focus on broad conceptual issues. The following chapter addresses itself to the broadest concepts we shall encounter in the book — to consideration of general systems theory and the fundamental characteristics of intellectual progress regardless of the specific field of inquiry.

Conceptual Problems

1. Successful executives develop an ability to sense symptoms of problems. Then they go on to solve the problems at the causal level. Less successful executives often work at the symptomatic level, never ferreting out the underlying causes of the problems. Although their actions temporarily cover the symptoms, the problems continue to recur. In designing a system, what types of sensors would you propose which would go beyond symptoms to underlying causes of problems? Give concrete examples of how such sensors would operate and what attributes would be measured.

2. Most firms follow the lead of one or a few firms in their respective industries. Select an industry in which you have a particular interest and consider which firms can be viewed as leaders and which are followers. What are the differences in their management styles, management systems, and key executives?

3. Assume that you are designing an inventory management system. How and for what purposes would you apply the following types of models: physical models, schematic models, deterministic mathematical models, and probabilistic models?

4. A schematic of the process of simulation is included in this chapter. Consider the problem of scheduling men and machines in production operations. Then demonstrate how this simulation process could be translated into concrete terms by testing various sets of decision rules. Then impact the system in such a manner that the model of the system must be modified. Having completed this, what modification, if any, would you make in the schematic of the process of simulation?

5. A procedure for the development of systems is discussed in this chapter. Using operations management as a frame of reference, consider the factors which should be included in the system conceptualization phase and in the model construction and simulation phase. Attempt to translate the concept into practice.

The Systems Concept

Introduction

3

Man's quest for knowledge has reshaped his world. From a seemingly random and chaotic environment he has fashioned a life experience in which he individually and collectively can affect and control his environment rather than remain subservient and responsive to it. This is especially true in the sciences and the many applied fields stemming from science. Knowledge of scientific laws and principles has allowed man to better understand phenomena and through this understanding to plan and control the behavior of phenomena.

Although great strides have been made in both the pure and applied sciences, lesser advances have been made in the social sciences. There are few laws in the social sciences which will withstand the rigorous scrutiny of laws in the physical sciences. The practice of business management is based on understanding of both physical and social sciences and thus understandably is viewed by many as a combination of both art and science. Until recently, emphasis has been placed on the development of sound business judgment, intuition, and feel for the situation as proper grounding for success in business decision making. In recent years, however, strides have been made, particularly in operations research, to apply scientific and mathematical methodologies to the solution of business problems.

Operations research and the mathematical models characteristic of it utilize a scientific approach to business decision making. And though they typically incorporate several variables and constants, mathematical models usually are severely constrained in terms of definitions of relationships and incorporation of all significant factors affecting the decision and factors which will be affected by the decision. It is this question of *relationships* which is central to the systems concept.

The utilization of the systems concept as an approach to the understanding and practice of management is becoming more important as the complexities of operating firms increase. Not only must a manager recognize the inputs with which he works and the outputs which he expects to achieve, but he must also know the information and physical systems which provide the transformation processes to convert the inputs into outputs. In light of this it may be useful to trace, abstractly for the moment, the fundamental questions to be answered in any inquiry into the nature and behavior of phenomena.

Man's quest for knowledge appears to have followed certain patterns regardless of the field of inquiry. When confronted with empirical phenomena which seemed random and chaotic his first task was to observe and understand the phenomena. He had to find answers to the questions *What* exists? and What are the characteristics of its static and dynamic existence? This led to descriptive studies of his environment. But knowledge of described phenomena is insufficient for purposes of prediction, planning, and control.

To reach this higher level of knowledge he had to ask a more difficult question: *Why* does the environment exhibit particular characteristics under varying conditions of its static and/or dynamic state? To find the answer to this question required the formulation of hypotheses about relationships among the described constants and variables. It required perceptions of the behavior of the phenomena under controlled conditions. In many cases it required experimentation to isolate the phenomenon in question and determine the magnitude of change and the causal factors affecting such change.

The search for answers to the question *why* has been long and arduous in many fields and, where successful, has led to theoretical constructs which explain not only what occurs but also why such events occur. In their highest form of development theoretical constructs provide not only understanding but also the ability to predict with certainty the results of purposeful modifications of the variables involved in the theory. Thus the predictive capability provided by a sound theory provides man with the tools to plan and control, in short to shape his environment and the conditions of his existence.

Invariably such theories have been based on perceptions of relationships among the variables and constants associated with empirical phenomena. They have explained the interaction of inputs of various types as they are transformed by natural or manipulated processes into predictable outputs. A set of such inputs, transformation processes, and outputs can be viewed as a system.

In many branches of science it has been discovered that several systems viewed originally as independent actually exhibit patterns of interaction and dependency with other systems. Conceptually at least, there are those who search for general systems which relate the systems within a branch of science; indeed there are some who go further by searching for the existence of general systems among the branches of science. Such interdisciplinary systems have been explored with useful results and some perceptive researchers believe it is conceivable there may exist a general theory of systems *per se.*

Boulding points out, however, that "general systems is a point of view rather than a body of doctrine."[1] As such it is a useful concept even though "at the moment it would be presumptuous to claim that there is any clearly defined body of theory which could be identified with the name 'general systems.' "[2]

Whether such a defined body of theory can be constructed is open to sharp debate. However the characteristics of the systems *point of view* are generally recognized and have been described by Boulding as follows:

1. The general systems proponent exhibits "a prejudice in favor of system, order, regularity, and nonrandomness . . . and a prejudice against chaos and randomness."

2. "The whole empirical world is more interesting when it is orderly. It is to the orderly segments of the world, therefore, that the general systems proponent is attracted."

3. If the general systems proponent embraces laws to explain order "he is ecstatic when he finds a law about laws."

4. He sets "high value on quantification and mathematization, for these are great helps in establishing order."

5. "Whereas the mathematician is content with the mere perception and demonstration of abstract order, the general systems man is interested in looking for empirical referents of these systems and laws of abstract order."

6. "The process of finding empirical referents to formal laws can easily take either one of two possible directions. We

---

[1] Kenneth E. Boulding, "General Systems as a Point of View," in M. D. Mesarovic, *Views on General Systems Theory* (New York: John Wiley & Sons, Inc., 1964), p. 25.
[2] *Ibid.*

may find some elegant relationship in the world of abstract mathematics and then look around the world of experience to see if we can find anything like it, or we may patiently piece out a rough empirical order in the world of experience and then look to the abstract world of mathematics to codify, simplify it, and relate it to other laws."[3]

In summary, systems as a point of view involves a belief in order, in relationships which are structured in terms of cause and effect. Even though the structural relationships may not be evident to a person with the systems point of view, he proceeds to inquire, experiment, relate, examine, and in other ways probe intellectually to find that elusive structure of relationships — the systemic construct.

In an analyst's attack on a problem he must first determine *what* exists concerning the phenomenon in question. He must measure the characteristics of the static and dynamic states of the phenomenon. He must not let his study of the nature of the phenomenon drift from empirical referents. He should avoid building on premises, principles, and precepts which are not based on observable facts. Only after he determines *what* exists is he in a position to move to the next stage of inquiry.

At the second stage he is faced with the question *why* the phenomenon exhibits particular characteristics under particular conditions. This path of inquiry may lead to the structuring of order in a systemic construct. To develop or discover such a construct is an arduous and time-consuming task. Its rewards, however, are well worth the effort.

The Systems Concept in the History of Science

One can more clearly grasp the systems concept by tracing its utilization in the quest for knowledge in a variety of scientific fields. In the areas of astronomy, biology, chemistry, physics, and engineering, systems concepts have been highly developed. As a result, man's capability to predict behavior of phenomena and thus, in many cases, to utilize his systematized knowledge for what he perceives to be useful ends has been greatly advanced. Certain social sciences, such as political science, sociology, psychology, and economics, also reflect the application of the systems concept. Here the progress has been slower but important strides forward are being made.

---

[3] *Ibid.*, pp. 26–30.

The fundamental question concerning us, however, is: Can the systems concept be applied effectively in understanding business phenomena and, in turn, can such a concept be used to predict, plan, and control business operations toward the achievement of organizational objectives? A brief survey of the development of understanding and knowledge in three sciences — astronomy, biology, and chemistry — will provide a historical perspective on this question. It will also demonstrate the absolute requirements of effective systems analysis:

1. The need to find out *what* exists in the real world — a commitment to empirical referents.
2. Measurement of the characteristics of the phenomenon in both its static and dynamic states.
3. Avoidance of building systemic constructs on premises, principles, precepts, and theories which are not based on observation.
4. The need to find out *why* the phenomenon behaves as it does in both the static and dynamic states.
5. Extreme care in classification of phenomena. Classification is a mental process and a human characteristic required by the limitation of the mind to relate and manipulate factors simultaneously. As will be seen, improper classification systems can be awesome roadblocks to understanding.

The Systems Concept and Astronomy

Armitage, in tracing the development of understanding of the science of astronomy, points out at the outset of his book that

> things do not happen in the world as if at the bidding of capricious sprites, but in an orderly manner, as if in deference to fixed laws which have to be obeyed. . . . The discovery of this essential orderliness of nature marks the beginning of what we call science. Once men have reached the scientific stage of development, they realize that success in living does not depend upon coaxing or forcing nature to do what we want. It depends upon understanding nature's laws, and in making use of them to serve human purposes.[4]
>
> Science also serves to satisfy certain needs of the human spirit. It helps us to understand the world and to feel at home in it. We are distressed by disorder, and always try to arrange in order the

---

[4]Angus Armitage, *The World of Copernicus* (New York: The New American Library, Inc., 1954), p. 15.

things with which we have to deal, whether they are the affairs of a nation, the books in a library, or our own ideas. Science satisfies us because it shows us that behind the transient and confused pageant of nature, there is a permanent and orderly reality. And it was in the sky that this order was first revealed to man on a vast and spectacular scale.[5]

Man's initial response to astronomical events was far from the response of one embracing the systems concept. He viewed eclipses and comets with awe and fear. They seemed to be signs of impending danger and the wrath of his celestial gods. Many cultures worshiped the sun and moon and would not dare to examine their characteristics as inanimate celestial bodies. Such an inquiry would have been heretical.

A very few took the first step toward knowledge by asking *What exists?* and What are the characteristics of its static and dynamic states? Their answer was based on the observation that the planet Earth appeared to be a flat circle which bisected the sphere of the sky bounded on all sides by the horizon. The sphere of the sky, which had the stars fixed to it, revolved slowly around the flat plane of Earth. The rotation of the sphere appeared to occur slightly faster than once per day. They observed that the moon, as viewed against the sphere of the sky, took about a month to make a complete circuit of the heavens; the sun completed its circuit in about a year. Thus calendars came to be based on these solar (annual) and lunar (monthly) cycles.

The invention of time as a conceptual system has had tremendous significance based on these initial and rather approximate observations of a year as 365¼ days composed of 12 months and 24-hour days each hour representing 60 minutes and each minute composed of 60 seconds. Until recently this system was so reliable that time was considered a constant against which variables in empirical phenomena could be measured. However, Einstein's Theory of Relativity has cast doubt that time is indeed a constant.

In passing it is also interesting to note that current business accounting practices are based largely on the monthly classification system or lunar cycles. This creates peculiar problems which have been solved by some firms through the use of thirteen periods of twenty-eight days each to minimize irregularities in monthly accounting periods. Nevertheless, for many other purposes, we still plan and control business activities according to the solar cycle. Fiscal years are often used to overcome this problem partially. At any rate

---

[5]*Ibid.*, pp. 15-16.

it is apparent what a peculiar classification system can lead to if carried on through the centuries.

Socrates, Plato, and Aristotle carried forward the ideas of an Earth-centered solar system which prevailed until the seventeenth century. The centuries of this traditional approach represented a period in which convention rather than innovation prevailed. The premise that all heavenly bodies move in circular orbits around a central Earth was not seriously challenged until Copernicus began to make his own observations of the movement of celestial bodies. He found that his observations did not coincide with the traditional system and therefore some elements of the prevailing system of astronomical behavior must be in error. With abundant faith in the belief that astronomy could be explained within a systems construct he challenged the wisdom of the ancients, the beliefs of his colleagues, and the overwhelming power of church and state. He dared suggest that Earth was merely another planet and that it and all other planets revolved around the Sun. This modification of theory provided for the development of a new system by which relatively accurate planetary tables could be generated, a more accurate calendar developed, and the behavior of planets predicted.

Copernicus' new system is discussed in *Six Books Concerning the Revolutions of the Heavenly Spheres*. It represents a lifetime of observation of what exists and formulation of a systemic construct so contradictory to prevailing thought that he did not consent to its publication until 1543, the year of his death.

Astronomers who followed the path of Copernicus were to refine his theories, always building their knowledge on a basic conviction of a *system* of relationships of heavenly bodies. Kepler (1571-1630) formulated his laws of planetary motion breaking the vise grip on circular motions to admit elliptical ones, as well as stating the relationship of the size of a planet's orbit with the time taken to go once around the orbit. His set of planetary tables superseded all earlier ones.

Galileo (1564-1642), whose observations of physical phenomena with the telescope revealed the universe as it had never before been seen, was finally able, through his observations of reality, to overthrow the Ptolemaic system in favor of the Copernican system as published in his *Dialogue Concerning the Two Chief Systems of the World*. His life exhibits remarkably strong conviction in a concept of systems in that he was twice brought before the Inquisition for holding his beliefs and ultimately was sentenced to perpetual imprisonment. The year 1642 marked his death and, in addition, the birth of Newton (1642-1727).

Although Newton's scientific contributions were much broader than astronomy, his work in the area concerning the understanding and prediction of planetary motion was far-reaching. Kepler's concept of elliptical orbits answered the first question, *What* is the behavior of the phenomenon in question? Newton, utilizing his concepts of the nature of gravitational force, provided the answer to the question, *Why* does this behavior occur?

In retrospect, the portrayal of this one idea in the science of astronomy indicates that in this field of science an abiding faith in the concept of a system or orderly set of relationships which explains phenomena and allows for prediction underlies our growth of knowledge in the field. Current programs based on space research follow these same channels. A system of relationships is assumed and the probes are designed to sense what exists and what are the characteristics of its static and dynamic states. In time we will have even better answers to the question *why*; and, as in the past, the systems concept will set the stage for such theories.

The Systems Concept and Biology

Perhaps the most profound development in the field of biology is Darwin's Theory of Evolution. A brief tracing of its roots and history gives some indication of the role of the systems concept in this field.

The effort to answer the question *What* are the characteristics of natural phenomena in terms of structure, behavior, and development? led to a notable system of classification developed by Linnaeus. All living things, he proposed, should be designated by two names—one indicating the genus and one indicating the species. His descriptive efforts resulted in elaborate systems of identification of both plants and animals with reference to their ancestral origins as follows:

Phylum
Subphylum
Class
Order
Family
Genus
Species

In essence, man's quest for knowledge in biology was greatly enhanced by creating a framework or system of classification which allowed him to view the body of knowledge in a systematic way rather than as the biological world appears at face value, that is, a random collection of dissociated plants and animals. Without such a systemic framework the mere description of what exists would have been most difficult.

The concept of a system in biology was therefore essential to answer the question *What* are the characteristics of the phenomena? The more difficult questions *Why* did species develop their characteristics? and *What* were the origins of species? were to shake the nineteenth and twentieth centuries no less than Copernicus' concept of the solar system shook the sixteenth century. The classic answer to this question was provided by Charles Darwin through his lifework and book the *Origin of Species*.

Predating Darwin were two notable theories on the origin of species. The *type* concept of Goethe was popular in its time and was built around the idea that each species arose from non-living matter or new matter as preconceived ideas. These preconceived ideas caused the specie to take its own particular form. The ideas yielding these forms were presumed to come from some universal mind. If ever there was a speculative, arm chair theory developed without reference to observation of phenomena, this was it. The acceptance of this theory rested in no small measure on the prevailing religious beliefs of the time with which it was generally in accord and was further bolstered by the fact that it bore resemblance to the Platonic concept of essences or abstract ideas existing in the universe and thus was in accord with the wisdom of the ancients.

The second notable theory was that of Count Louis Buffon (1707-1788), a naturalist, whose thoughts on the origin of species rested on observation of phenomena rather than on speculation dissociated from reality. He first posed the proposition of evolution in his studies of the species *Homo sapiens* in 1745. His first premise, that all human races can interbreed successfully, met the test of uniformity of the species and his second premise was based on the Biblical reference to a common pair of ancestors, Adam and Eve, from whom all humans are descended.

Buffon's observation of remarkable differences among human beings left him with no other conclusion than that these differences resulted from the effects of the environment on the species. He expanded his ideas to suggest further that all animals must have *descended* from original archetypes of their species and their present dissimilarities were due to environmental influences.

As a religious man in an age of absolute authority of the church, Buffon, like Galileo, was taken to task for his ideas and condemned by the church and his colleagues at the University of Paris. His statement stands as vivid testimony to the powers of convention over innovation: "I declare that I had no intention to contradict the text of Scripture; that I believe most firmly all therein related about creation, both as to order of time and matter of fact; I abandon everything in my book respecting the formation of the Earth and generally all that may be contrary to the narration of Moses." This censure by the church occurred even though Buffon, so close to the concept of a system of evolution, based his thinking on Biblical referents rather than empirical referents by suggesting that present species have *degenerated* from the perfect archetypes created originally by God.

Charles Darwin (1809-1882) moved the theory of evolution forward by countering Buffon's proposition of degenerating species with the suggestion that the species improve and adapt to their environments. Indeed, failure to improve and adapt leads to extinction. Thus current representatives of the various species represent evolved adaptations of earlier species.

Before Darwin published his answer to the question *why* in the *Origin of Species* in 1859, he had spent a lifetime observing *what* are the characteristics of the phenomena in question, sparked by his five-year round-the-world voyage on the H.M.S. Beagle initiated at the age of 22. These years of observation produced a preponderance of evidence in favor of natural selection which could not be ignored. The battle which subsequently was waged between Darwinists and those embracing the concepts of Judaism and Christianity related to the origin of species and particularly the origin of man has had long lasting repercussions. In some sectarian schools and colleges the teaching of evolution is still banned.

It has been suggested that a systemic construct is the key required to unlock the door of understanding. To describe isolated phenomena is useful, but perception of systemic relationships among the phenomena in terms of inputs, transformation processes, and outputs is the catalyst for reaching advancements in knowledge. Darwin's unflagging efforts at description based on empirical referents and his keen perception of a systemic construct with reference to transformation processes ably illustrates this proposition.

The Systems Concept and Chemistry

The study of chemistry, like that of biology and astronomy, began early in the history of man. As was the case in the other sciences we have discussed, the first observers were prone to speculate about phenomena without utilizing empirical referents. The belief in reason as the road to truth led to many convictions which persisted for many centuries about the basic chemical elements.

One of these convictions was that four elements were fundamental: earth, air, fire, and water. For centuries alchemists attempted to turn lesser metals into gold and spent their time applying incomplete knowledge or incorrect principles in attempts to produce desired results. Only in the last century or two has alchemy given way to the science of chemistry.

The transition was possible only because these early chemists realized that conceptual referents were insufficient for understanding. To make the necessary breakthrough they set about to find out *what* exists in terms of the particular phenomenon in question. They carefully experimented, under controlled conditions, to determine the characteristics of matter in static and dynamic states. They also tested relationships of different types of matter to determine what in fact exists.

After turning away from conceptual principles which were faulty they turned their attention to developing an understanding of what exists. With that understanding they were then in a position to cautiously conceptualize and theorize to find the answers to the question *why* matter is as it is and reacts as it does. Although there are many examples of systemic constructs in chemistry, one stands out as a clear example of the systems concept in this field. That systemic construct is known as the periodic system of the elements.

The development of the periodic system or table of the elements was a slow process, beginning in 1817 when J. W. Dobereiner noticed similarities in certain triads of elements, notably calcium, strontium, and barium; chlorine, bromine, and iodine; and finally lithium sodium, and potassium. By 1854 other elements had been added to these triads; fluorine to chlorine, bromine, and iodine; magnesium to calcium, strontium, and barium; and other sets were established based on their observed characteristics. Oxygen, sulfur, selenium, and tellurium were grouped together and nitrogen, phosphorus, arsenic, antimony, and bismuth fell into one classification.

In 1862, A. E. B. de Chancourtois developed a set of relationships wherein the difference of 16 in atomic weight appeared to have significance. A year later, J. A. R. Newlands proposed a system of seven groups with seven elements in each. Finally a breakthrough was made in terms of a useful systemic construct when D. I. Mendeleev developed his periodic system in 1869. After making careful experiments, Mendeleev determined that his systemic construct was essentially correct but to his dismay he found that the atomic weights of several elements did not fit the system precisely.

With faith in the concept of system he proclaimed that these atomic weights must be wrong. As was the case with Copernicus, who found the planetary tables not in conjunction wth ordered systemic relationships, Mendeleev and other chemists proceeded to examine these discrepancies and found that errors had been made in observation and experimentation and that the system provided a clearer insight into reality than isolated examination of individual elements.

Even though Mendeleev's work was somewhat crude by modern standards, his systemic construct allowed him to structure a system of elements with which he was able to make some astounding predictions. Perhaps most interesting by modern standards was his prediction of six elements which he said existed even though they had not been discovered. They fit into his system yet had not been found experimentally. He called them eka-boron, eka-aluminum, eka-silicon, eka-manganese, dvi-manganese, and eka-tantalum. Soon afterwards the first three of these were discovered: scandium, gallium, and germanium, respectively. Later the other three, technetium, rhenium, and polonium, were found. All had characteristics closely related to those predicted by Mendeleev.

Following the discovery of helium and argon, the periodic system indicated the existence of four related elements: neon, krypton, xenon, and radon. In time the existence of other elements was predicted in a similar manner. A comparison of Mendeleev's predictions concerning the properties of germanium with its properties after discovery is a striking example of the capability of a systemic construct to expand the frontiers of knowledge. On many technical points he was surprisingly accurate.

It is interesting to note that although Mendeleev and others working in the area of building a periodic system of the elements came up with a system capable of prediction and valuable for analysis, they essentially "pieced out a rough empirical order in the world of experience," as Boulding puts it. The real understanding of the structure of the periodic system in terms of *why* the relationship is as it is did not come until much later with studies of nuclear structure.

While Mendeleev and his colleagues were concerned with relationships among atomic numbers, atomic weights, and valences, the underlying explanation of the periodic system rested on characteristics of matter much smaller than the atom. The system can now be explained in terms of the basic structure of atoms, especially their capacity for electrons. Without delving into the theory of atomic structure, it has been found that atoms contain electron shells which have limited capacity. When this capacity is filled, an outer shell is required and so on. The capacity of these shells appears to be the determinant of the periodicity of the system and at this point explains why the periodic system exists.

We have briefly traced three significant insights in the fields of chemistry, biology, and astronomy. Each rests on the assumption of an underlying system of relationships. The discovery of these systems required analysis of what in fact exists rather than on further development of subjective speculation about the phenomenon in question. In each case former premises, principles, precepts, and theories had to be abandoned and their abandonment was accompanied by strong antagonism toward a new viewpoint. In each case the variables and constants involved required identification based on empirical referents. In each case the characteristics of the phenomenon were measured and relationships determined. In each brief summary of developments in these fields, the hazards of improper classification became apparent. After classification systems had been brought in line with what exists according to empirical referents, after measurements had been made of the variables and constants associated with the phenomenon, and after systemic relationships had been established, then the way was open to answer the question *why*. The answers which followed served to advance substantially the knowledge in these fields.

The Systems Concept and Medicine

There are many advances in knowledge in the field of medicine which emphasize the points made above. Each runs parallel to the histories cited in astronomy, biology, and chemistry: the early reliance on superstition; rationalization and theorizing based on subjective assumptions; the men who doubted these theories and placed themselves in jeopardy by dissecting cadavers illegally; those who spent their lives carefully documenting the characteristics of phenomena; and finally those who answered the question *why*. This process, of course, still goes on today. But in medicine we start with experimentation and laboratory analysis and derive conclusions from that.

An important analogy can be drawn from medicine to give us an insight into possibilities for the future in business management. The human body was regarded by most early observers to be made up simply of bones, muscles, various organs, and other assorted features. No real progress was made in understanding the body until various researchers began to view it as an integrated set of systems. The conception of a circulatory system and the organs, tissues, distribution patterns, and blood which are a part of it made sense only when viewed systemically rather than as separate things.

The concept of other systems likewise provided important insights. Among these systems are a skeletal system, a muscular system, a respiratory system, a nervous system, a reproductive system, a digestive system, and an endocrine system. Physicians of course can identify dozens of other systems and subsystems. The point is this: until a systems approach to medicine became the research methodology, the understanding of medical phenomena was severely limited.

Today, with understanding of individual systems and the interactions among systems, physicians and surgeons are able to accomplish two important things which have yet to be accomplished in business management. First, they have become highly skilled in diagnosis. That is, they can observe surface manifestations of a physical problem and, through knowledge of systemic relationships and objective tests of the manifestations, they can work back from results to the basic causes. All too often in business we spend our time treating symptoms or, more colloquially, "putting out brush fires." Unless we understand the systems and subsystems which exist in the business firm and the relationship of the firm to its environment, and until we develop objective tests of the manifestations of problems, we will be faced with limited diagnostic ability in solving business problems. Too often we must settle for the treatment of symptoms rather than the correction of fundamental causes of the problems. Without a systemic construct of the firm it is most difficult to trace back to causal factors. And unless causal factors are treated, the problems will recur.

The second fundamental accomplishment in medicine which has yet to be achieved in business concerns the ability to introduce a specified change in the system and be able to predict a specific result. A physician follows his diagnostic work with a plan for corrective action. In most cases this involves introducing some type of chemical compound, microorganism, physical change, or surgical procedure. In each case he knows with a rather high degree of certainty which changes or stimuli to introduce into the appropriate subsystem. He is aware of the magnitudes of these changes which are appropriate to solve the problem at hand. Finally he knows with some certainty what the results will be, and their achievement provides feedback information. Should crises develop, emergency procedures have been developed to sustain the system until home-ostasis is again established in a healthy organism. Without being grim about it, he also knows when the system's condition is terminal and when further resources will not result in its survival.

Contrast this ability of the medical doctor to that of the business manager. His diagnostic skills are limited and he therefore attacks symptoms of problems more often than causes. When prescribing a corrective course of action, he has limited understanding of what a correct course of action should be. This is quite natural since there is limited knowledge of the systems of which the firm is composed. In addition, for lack of experimental data on the measurement of variables and constants and their interaction patterns, he has limited knowledge as to how much corrective action to apply. Con-sider price changes or budget cuts as cases in point. Magnitudes are of critical importance but there is a limited store of knowledge to be drawn upon to know the proper amounts to achieve desired results. As a matter of fact, once the course of action has been chosen and the decision made as to magnitude of change to introduce in the system, there is still no assurance that desired results will be obtained. The business manager's ability to predict the conse-quences of these actions can hardly be compared to that of the physician. Again, lack of knowledge of the stream of cause-event sequences throughout the system accounts for the uncertainty and increases the risks.

To continue the analogy with respect to application of courses of corrective action, it is common to find physicians backed up by emergency procedures to sustain the system. In the business firm such procedures are not highly developed. There are no oxygen tents, no hypodermic syringes ready with adrenalin and other potent drugs, no surgeons standing by for radical modification of systemic relationships. It is not so much that emergency procedures could not be developed as that they have not been carefully studied. In business crises, post mortems often indicate that too little correc-tive action was applied too late.

This analogy is not presented in order to compare business man-agers unfavorably with physicians. In terms of knowledge, skill, and other attributes these two groups of men are both talented and able. It is intended rather to point out how much more effective business managers could be in diagnostic skills and the application

of corrective courses of action — in short, the practice of management — if they had the knowledge of their firms that physicians
have of their patients. In medicine this knowledge has come from
observation of the phenomenon in question, the human organism,
using empirical referents. It has come from careful examination of
its characteristics in both its static and dynamic states. Further, it
has come from analysis of measurements of changes in the phenomenon over time and as the systems are affected by external and
internal stimuli. And most importantly, it has come from the view
that the human organism can best be studied and understood as a
series of systems and subsystems.

There is reason to believe that the systems concept, which has
proved greatly beneficial in medicine and other fields, may be of
equal value in business. The road to a useful systemic construct for
business firms which will meet pragmatic tests will be a long and
arduous one. There are, however, those in business firms and in
academia who are dedicated to the investigation of business systems, and although their concepts and experiments will not lead to
a unified theory for many, many years, each probing effort is useful
as an effort to push back the frontiers of knowledge in the complex
field of business management.

Summary of the Systems Concept

The exploration of the use of the systems concept in this chapter
focuses on its effective use in bringing about major breakthroughs
in knowledge in a variety of fields. The following selected cases
serve to emphasize this point: 1) the development of the Copernican
heliocentric theory in astronomy, 2) the development of the Darwinian theory of evolution in biology, and 3) the development of
the periodic table of the elements in chemistry.

In each case, two fundamental questions provide the framework
within which the researchers worked. First: *What* exists and what
are the characteristics of its static and dynamic states? Second: *Why*
does the phenomenon in question exhibit particular characteristics
under varying conditions of its static and/or dynamic state? The
finding of answers to these two questions yields theories which provide not only understanding but also the ability to predict the results of purposeful modification of the variables involved in the
theory. Such predictive capability is required for effective planning,
analysis, and control in the area of operations management.

The historical perspective provided in this chapter yields certain absolute requirements for effective systems analysis which form the basis of more detailed patterns of theoretical development later in the book:

1. The need to find out *what* exists in the real world — a commitment to empirical referents.
2. Measurement of the characteristics of the phenomenon in both its static and dynamic states.
3. Avoidance of building systemic constructs on premises, principles, precepts, and theories which are not based on observation.
4. The need to find out why the phenomenon behaves as it does in both the static and dynamic states.
5. Extreme care in classification of phenomena.

A comparison of the practice of medicine and the practice of management indicates that in a least two important respects physicians lead managers: first, in the development of diagnostic skills; and second, in the area of the application of corrective courses of action. It is proposed that utilization of the systems concept in operations management will tend to close this gap.

At this point we have explored some of the fundamental ideas at the general systems concept level depicted in Figure 1-1. Our next level of concern in the exploration of systems analysis in operations management is the next level in the systemic hierarchy in Figure 1-1 — the environmental systemic construct. At this level we shall explore systemic interfaces of the firm with the objects in its environment: government, customers, competitive firms, the community, labor unions, stockholders, banks, and suppliers.

Conceptual Problems

1. Operations research and the mathematical models characteristic of it utilize a scientific approach to business decision making and, though they typically incorporate several variables and constants, mathematical models usually are severely constrained in terms of definitions of relationships and incorporation of all significant factors affecting the decision and factors which will be affected by the decision. How could such factors be incorporated to achieve more useful decisions where operations research techniques are used?

2. Boulding in his article "General Systems as a Point of View" suggests that general systems is simply a point of view or way of looking

at the relationships among empirical phenomena. Other proponents of general systems theory suggest that general systems do in fact exist and that the discovery of truth with respect to the laws of nature and other statements with respect to the relationships among empirical phenomena rest on the discovery of the fundamental systemic construct underlying it. Which position would you support and why?

3. It has been stressed that effective systems analysis requires among other things a commitment to empirical referents. In other words the analyst must determine what actually exists rather than base his analysis on symbolic constructs of what exists. How can an analyst achieve this in large complex organizations where the bulk of his perceptions are normally drawn from information systems rather than physical systems?

4. It has been suggested that a systems analyst avoid building systemic constructs on premises, principles, precepts, and theories which are not based on observation. How does one determine whether existing business premises, principles, precepts, and theories are based on observation? In what ways can they be tested empirically for validity?

5. Improper classification systems can be awesome roadblocks to understanding. What are some examples of this in science and business? How can they be modified?

6. What are some of the striking similarities in the history of science with respect to the use of the systems concept and its acceptance? Will history repeat itself with respect to the application of the systems concept in operations management? Why?

7. Contrast the diagnostic capabilities of the physician to those of the business executive. In what ways do they differ? What are the ways in which the diagnostic capabilities of the businessman could be improved using concepts drawn from medicine?

The Environmental
Systemic Construct

The Environmental Set

4

A system can be defined as a set of objects with relationships between the objects and between their attributes. For purposes of system design and systems analysis, the system must have objects or components capable of being defined and identified. These components or objects must have attributes or characteristics which can be sensed and, it is hoped, measured. Finally there must be relationships among the objects and their attributes; in short, we must know how they are linked together.

A system could be viewed as a set of static objects with static attributes and static relationships, as is reflected in most organization charts and other schematics of organizational structure, but this provides a point of departure of only limited usefulness in the examination of a business firm which has changing objects with dynamic attributes and ever-shifting relationships. To develop a feel for a dynamic system, it is useful to view the operation of the system as a series of inputs which undergo transformation processes yielding outputs. Put into a simple schematic, the operation of the system could be depicted as follows:

Inputs ——➤ Transformation Processes ——➤ Outputs

In order to develop a systemic construct for business firms it would appear necessary that objects, attributes, and relationships be defined and that, in a dynamic sense, an input-output relationship exist among the objects with useful transformation processes taking place. Although any classification system is subject to error, a tentative one could be structured around organizations, particularly with reference to the environmental set within which the firm functions. A tentative set of objects and relationships at the environmental level is proposed in the schematic in Figure 4-1.

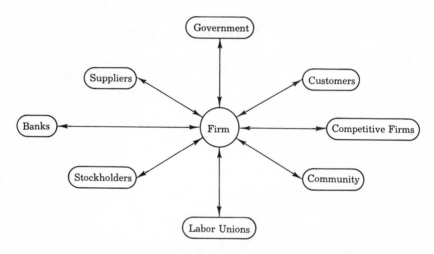

Figure 4-1.   The Environmental Set of the Firm

If the firm is viewed as the system within an environment, then the objects or entities in that environment might be denoted as suppliers, banks, stockholders, labor unions, community, competitive firms, customers, and government. In each case there is little question that the firm is linked to each of these environmental objects, although the relationships may not be tightly defined.

In order to utilize systems analysis for management and decision making, one of the critical dimensions in terms of usefulness is the precision with which such relationships can be defined. In the physical sciences such relationships can be defined and controlled under laboratory conditions. Environmental temperature and pressure controls are basic to most laboratory phenomena as well as the ability to isolate environmental influences and test the relationships one at a time. In the case of business applications these characteristics of laboratory analysis are rare. Occasionally management can isolate an environmental factor and measure its impact on the system and the impact of the system on the environmental factor, as in certain restricted tests in marketing research. However, in virtually all business circumstances, the interaction of environmental factors with the system and among the factors is subject to processes not fully understood.

The development of understanding in the environmental area presents a great variety of challenges that will require a great deal of future research. Fortunately, these challenges are being accepted by universities as interdisciplinary teams attack the problems. They are also being met in industry as more and more firms engage in management research in addition to their traditional product and process research activities.

An important added dimension of the environmental problem concerns the fact that each of the objects in the environment of the firm can be viewed as a system in itself. As such it interfaces with virtually all the other objects to some degree. The complexity of these relationships is depicted in Figure 4-2. In order to grasp the current state of the art with respect to the objects in the environmental set, we shall examine each of the interfaces.

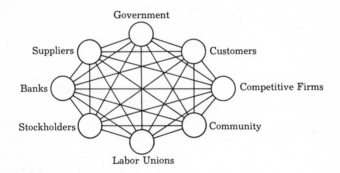

Figure 4-2. The Environmental Set of the Firm as a Set of Interacting Systems

The Customer-Firm Interface

In the case of customers we know that the fundamental linkage with the firm is in terms of an output stream of products and/or services which meet the perceived wants and needs of the customers. The basic input-transformation process-output concept is one of transforming materials, manpower, financial resources, energy, and knowledge into marketable products and services. This tends to be a one-way process from inputs through the production-distribution processes to the customers, with some provision for returns.

There is also a vital input-transformation process-output stream from customers to the firm in terms of sales dollars, credit income, and information with respect to the products and/or services rendered. Most firms devote a good deal of effort to this critical firm-customer interface. In a way it is the fundamental open-loop connective between the firm and its environment.

Much has been written on economics, marketing, advertising, pricing, distribution, and other aspects of this interface, yet consumer behavior remains a most puzzling phenomenon. Even such aggregate measures as projected national sales of a given product (automobiles, refrigerators, or color television sets, for example) are subject to considerable error.

Whereas an understanding of customer behavior used to be based entirely on subjective opinion, there has been a strong tendency in recent years to incorporate analytical quantitative models in this area. Marketing research of the past two decades has provided many answers to the question of *what* exists in this interface. Numerous statistics on sales levels, manipulated in every conceivable way through descriptive statistics and countless surveys, have created a large backlog of data about the interface. In many ways we know the answers to what the consumer has done, how he has behaved in a given set of circumstances at a given point in time, where and when he acted. With respect to the question *why,* much challenging research remains to be undertaken.

Why do consumers behave as they do? Motivation research, developed during the late 1950's, was designed to answer this question. Given a certain product, the motivation researchers attempted to find the answer to the *why* question by applying depth interviews, sentence completion tests, word association tests, thematic apperception tests, story completion tests, Rorschach inkblot tests, and other tools borrowed from the field of psychology. This provided new and unique insights concerning consumer behavior, yet even these techniques have proved insufficient to the task of explaining the customer-firm interface.

Currently, the tools of operations research are being applied in marketing, and computers are being used to carry out analytical statistical studies of marketing phenomena. The operations research experiments, although still inconclusive, are being applied in terms of looking for answers to the question *why* using measurable stimuli affecting the firm and its interface with its customers. The transition from descriptive statistics to analytical uses of statistics in marketing should provide further insights in the future.

If the understanding of consumer behavior is still incomplete in terms of the consumer's relationship to the firm, we might hope that other environmental interfaces with the consumer would clarify the situation. Unfortunately, the situation is even more ill-defined. Most firms have spent a considerable amount of time and money to develop an understanding of relationships with "their" customers. Yet much less effort has been expended on and fewer results obtained from the study of the relationship of customers to systemic impulses generated from and received by competitive firms, labor unions, banks, and the government.

It seems clear that significant relationships exist among these environmental objects and they, in turn, have an effect on the customer-firm interface. The challenges to researchers and practitioners in this area should provide opportunities for important interdisciplinary efforts. When the results of research in this area become available, then the application of systems analysis to the solution of business problems will yield improved results. Progress is being made continually and current research, building on the marketing research of the past, both objective and subjective, should lead in time to a clearer understanding of the firm and its customers.

Once the significant variables are identified, measurements gathered, and correct relationships discovered, the stage will be set for effective use of systems analysis at the environmental level. Until then we can expect to see a good deal of subjective experience, intuition, and judgment utilized in making important marketing decisions.

The Competitive Firm-Firm Interface

It seems apparent that in a systemic sense, in a free-enterprise system, the impulses of the firm on its competitors and those of competitive firms among themselves affect the state of the firms involved. Price changes, trade-offs in share of market, advertising strategies, and product mix modifications are all manifestations of this interaction.

To utilize systems analysis with respect to the interface of the firm with its competitors, it is necessary to determine *what* exists and *why* the system behaves as it does with respect to this environmental component. Finding answers to the question *What* exists?

in terms of the competitive interface is often more difficult than in the case of the customer-firm interface. Most firms know what they *have done* under the pressures of competitors in the past. Some firms know what they *will do* in the future in the light of new pressures exerted by competition. But only a few firms *act instead of react* to competition. These few usually set the pace as leaders in their industries.

Thus it seems in a systemic sense that the inter-firm actions of a few trigger reactions from many. If one could find data on these few firms and measure the magnitude of their key actions and follow through time the effects and counter forces exerted by their competitors, one might be able to piece out the semblance of system objects and relationships.

Although we think we know which firms are objects in the system, i.e., the competitors of the firm in question, recent developments with merging companies and technological breakthroughs leading to substitute competition have added new dimensions to this question. Where, in the past, competitors developed "workable" competition among themselves, it appears that the concept of competitive firms is much broader and more elusive than it has been in prior years.

In addition to this development, an understanding of the competitive firm-firm interface will require access to accurate and pertinent data concerning the actions of competitors in the past and present, and especially, proposed actions in the future. Even at that point in time when firms are in a position to provide data on their competitive actions, it will be some time before researchers can find the answer to the question *What* exists? with respect to this interface. Only after such data are gathered and analyzed will researchers be able to attack the more important question, *Why* do competitive firms behave as they do in a dynamic environment? Any answers to this question without recourse to factual data from the firms involved must be limited to subjective speculation. And as was pointed out earlier, our lessons from the use of the systems concept in science seem to indicate that such subjective speculation without reference to empirical phenomena often leads to incorrect understanding of relationships.

The Community-Firm Interface

There was a time when the firm was the community, or so the founders thought. Some colorful history has been written about company towns, where the company owned not only the plant but also the homes of the workers, the stores, schools, hospitals, fire department, water works, and recreation facilities. In most such histories, usually written for the company by the company, the hero is the founder. His picture is painted in soft, humanistic pastels and he comes across not so much an entrepreneur as a philanthropist.

Few such histories have been written in the past thirty years. Accounts of such micro "industrial democracies" have been reported but somehow the pastel shades have faded and "humanistic" and "philanthropic" are hardly the terms used by current authors. In retrospect, they often see the founder as sort of a latter-day feudal baron or outright socialist. Why?

The basic sentiment involved reached a high-water mark when Charles Wilson uttered his famous (or perhaps infamous) words: "What is good for General Motors is good for the country." If one carefully considers the economic side of this argument, it appears that Mr. Wilson was quite correct. Yet members of communities, citizens of the United States, felt at the time that perhaps the truth lay closer to the reverse statement: "What is good for the country is good for General Motors." They believed that there were other standards, more important than those of business, and other goals, more important than profits.

Such sentiments indicate that the objectives of the firm may not be congruent with those of the objects in its environmental set. Clearly the value systems vary to some degree. Analysis of these differences should provide a better understanding of goal formation processes and improve supportive relationships between the firm and the objects in its environment.

During the 1950's and early 1960's business, government, and labor grew as institutions maintaining homeostasis within the state by countervailing power according to Galbraith. The common man seemed to shrink in comparison. Yet in the last analysis, in a democracy the power remains with the people, and in recent years this power has been exerted in many ways.

Urban problems, civil rights issues, pollution, and many other areas have been pressed to the point where business, labor, and government have had to reassess their positions. The demands of the people for clean air, clean water, truth in lending and packaging, automobile safety, and other demands have been laid before businessmen. In brief, the position of the citizens has been that business firms must meet their community responsibilities, as defined by the citizens not the firms, or else businessmen would have their freedoms of courses of action in the conduct of business impaired. This represents a drastic departure from the company-town days.

Many firms are now heeding this message. They have engaged their personnel in civic affairs to solve urban problems, they have attempted to meet the terms of civil rights, they have started on the problems of pollution, and they find themselves constrained by law on other problems concerning which they were too slow to act and reaped restrictive legislation for their inactivity. Yet there are still other firms — slow to respond — which are operated as if the community were a satellite appendage.

In a systems sense, the community-firm interface is quite ill-defined. The objects of the system appear to be the firm and the community. Although defining what is meant by the *firm* is not difficult, it is very difficult to define what is meant by the *community*. As to the question of which attributes to measure concerning these objects, the challenge to researchers is significant. Shall it be the amount of money the firm contributes in taxes, or to the United Fund, or to the building of the community swimming pool? These measures seem insufficient when weighed against the community problems recently left on the corporate doorstep.

In general, it can be concluded that finding the answer to *what* the correct attributes should be, or *what* indeed one key object in the environment — the community — really is, represents a substantial and pressing research challenge. As to the relationships between the community and the firm, there seems to be even more uncertainty. As in the case of the competitive firm interface with the firm, it seems here as well that our hope for understanding the answer to the question *why* must be based on much future research in this area. That research is just beginning. With luck a few answers will be found in time to solve some of these problems before the environmental object, the community, forms alliances with other environmental objects such as labor unions and government, to the detriment of the firm.

The application of systems analysis to this interface, in an effective manner, lies well in the future. Of all the environmental objects cited, this one requires the most pioneering effort. Fortunately, a substantial national commitment has been made to the study of communities. Much of this research should spin off significant insights into the nature of the community-firm interface.

The Labor Union-Firm Interface

In the case of labor unions it is apparent that the fundamental linkage with the firm is one of an output stream of wages, salaries, and fringe benefits from the firm to the employees, under conditions negotiated with the unions. The opposing input stream involves the provision of labor from the employees to the firm. Labor unions represent a buffer organization between the employees and the firm.

In a systems sense, manpower can be viewed as one of the significant inputs which, when accompanied by other inputs, is instrumental in producing useful outputs. Although employees in certain categories, such as managers or engineers, are not usually viewed as environmental variables outside the firm, it is becoming clearer that most blue collar and many white collar employees can be viewed as environmental variables when they are represented by labor unions. They often pledge their allegiance to the union rather than to the firm.

Our understanding of *what* exists with respect to the firm-labor union interface is colored by emotional responses which have been built up over the last century. From a systems point of view, it would seem that the labor input to the firm is a matter of resource allocation and the trade-off in funds to pay for the resource would be based on market prices for the labor. This relationship would be analogous to the trade-off involved in other resource areas such as raw materials, supplies, equipment, and electrical power. In these cases, a limited amount of negotiation is involved within fairly limited parameters based on market conditions. However, with respect to the firm-labor union interface, negotiations play a much larger role and competitive relationships and market prices a much smaller role.

In a sense, the relationship between the firm and the labor unions with which it deals revolves around the resolution of conflict in which relative power has the dominant role rather than rational analysis of economic contributions and rewards. Even in highly integrated systems some conflict resolution problems exist, but in most cases the trade-off decisions are based on analysis of costs and benefits. It is difficult to perceive of the utilization of a technique such as value analysis in labor negotiations even though value analysis represents a significant stride forward in the acquisition of other resource inputs to the firm.

A brief review of historical developments in the labor area may shed some light on this phenomenon. The labor union, as an institution, developed from felt needs for representation by workers in dealing with management. Undoubtedly, several decades ago, management manipulation and exploitation of workers existed. Subsistence wages were paid, job security was tenuous, working conditions were poor and in many cases unsafe.

With the advent of government intervention in the management-labor interface in the 1930's, the balance of power was shifted resulting in minimum wages, a higher degree of job security, and improved working conditions. In recent years some evidence has accumulated that the balance of power has continued to shift in favor of labor.

The manifestations of this shift in power have resulted in some unusual consequences affecting both management and labor which systemically tend toward disequilibrium. For example, rapidly rising wages have created a situation in which some firms have gone out of business and the workers lost their jobs instead of improving their economic status. In other cases, increased wage levels and fringe benefits have made automation a more economic alternative to manpower, resulting in lost jobs. Still another case involves reduction in the work week, presumably to provide more leisure time. It was discovered that many of these workers consequently moonlighted, taking two jobs involving more work hours than they worked earlier.

The wage-price spiral is another indication of serious system disequilibrium in the aggregate sense. Provision for ultimate job security has created conditions where the size of the workforce cannot be effectively reduced, resulting in inefficient operation of the firm. A common consequence of conflict is the use of the strike to put pressure on management. In the past this has resulted in lost pay for the workers and lost production for the firm. In many cases these losses are not recoverable by the firm or by the employees even though the employees receive increased wages. Recently this sort of disequilibrium has been intensified when firms take the stand that a strike against one is a strike against all in their industry. Such firms retaliate by closing plants that are not being struck and even more productivity and employee wages are lost, not to mention the adverse impact on the public.

All in all, the firm-labor union interface is similar to that of countries at war when resolution of conflict is at hand. Each has its arsenal of weapons, its strategies, its spies and counterspies, and on the sidelines stand the members of the U.N. Security Council, so to speak, the concilliators, mediators, and arbitrators. It is clear to most people that nobody wins a war between nations; it seems to many that the industrial-labor war is also a no-win situation.

Management contends that it always loses something in these conflict situations. Labor unions, to justify their existence, must gain something from management in terms of wages, fringe benefits, and security in return for the workers' dues. So labor must make demands on management, periodically engaging in a major scheduled confrontation while keeping the pot boiling between confrontations with minor skirmishes about grievances with the resultant walkouts, slowdowns, and so forth.

Effective use of systems analysis should provide a reasonable alternative if it centers on the inherent symbiotic relationship between management and labor. If the relationships as now established continue to operate in the future as in the past, major system disequilibrium for the firm and even the economy will result. It has been noted that those who do not study the past are destined to relive it. Unfortunately it appears that the organization efforts being expended by labor unions in the areas of white collar workers, teachers, farmers, and government workers are patterned after the industrial experiences of the past. Similarly, the reaction of administrators and the public is much the same as it was in the 1930's. The unfortunate conclusion one could draw from this is that system disequilibrium will continue in the future and on a much broader scale.

There is no question that a great deal of data have been gathered on the management-labor interface. Yet our understanding of what exists and why is insufficient to remedy obvious deficiencies. It is true that certain experiments have been made to alter the fundamental relationships by starting with other premises such as profit-sharing plans and guaranteed annual wages. Yet these have not changed the course of events to any large degree.

The concept of an integrated, balanced system operating over time in relative equilibrium is one which appeals to management. Labor unions have somewhat different objectives based on another value system. Management's and labor's inability to reach relative equilibrium rests to a large extent on starting with the wrong premises and utilizing military models to achieve this balance. In order to achieve desired results for all parties involved, much more study must be made of the systemic relationships between them. Management and labor are in reality supportive of one another and through understanding of their positive and reinforcing relationships it may be possible in the future to achieve effective symbiosis.

In addition to the complexities of the management-labor interface are other complicating relationships. Government intervention is a significant factor in terms of legislation, mediation, and arbitration. Various rulings affect management, labor, or both. Customers, another environmental object, react to potential labor problems associated with the firm in many ways leading to imbalance in the system. For example, forward buying prior to a possible strike increases customer inventories to a point where sharp sales declines will occur in the firm if the strike does not take place. The same kind of imbalance may exist between the firm and its suppliers when they are threatened by strikes. Unfortunately, these inventory imbalances require time-consuming oscillation before conditions return to a state of equilibrium.

As with the labor union-firm interface, interfaces of other environmental objects as they affect both management and labor provide unique research challenges to determine *what* exists systemically. Even more challenging is the study of *why* an impact on the firm at one point will yield given outputs at other points. Mere description and analysis, however, will not be sufficient for effective systems work in this area. To ultimately incorporate the environmental object, labor, as represented by labor unions, into a balanced system capable of serving the needs of all objects in the system and its environment, some fundamental changes must be made in the premises and practices of labor-management relations. Challenging and innovative work remains to be done to discover what the requisite changes should be.

The Stockholder-Firm Interface

The fundamental linkage between the firm and its stockholders involves a flow of money into the firm and a two-fold flow from the firm to the stockholders in terms of dividends as a return on their investment and growth in the value of the stock for long-term gains. A secondary linkage involves the voting rights of stockholders which, according to theory more often than fact, give the stockholders ultimate control over the management of the firm.

Much data have been gathered and analyzed to determine *what* the nature of the stockholder-firm relationship is and *why* stockholders and the firm behave as they do, given particular modifications in systemic relationships. The backgrounds and temperaments of financial analysts, in both the firm and investing institutions, tend to be such that cause-effect relationships are investigated and the attributes associated with these relationships are measured and manipulated quantitatively. In terms of the money inflow and dividend outflow pattern, enough data have been gathered to allow for meaningful systems work in this area to provide the necessary balance for both the firm and its stockholders.

In terms of the secondary linkage, voting control of the firm, much more research would be beneficial. In the past, stock ownership, whether in small or large firms, meant control of management because the major stockholders, in addition to having voting rights, were the managers of firms. With the development of stock markets involved in mass merchandising to small investors, voting power has been tremendously diluted. Indeed, most small investors have no interest whatever in controlling management or taking on active management responsibilities. They look on their ownership share simply as an investment, much as they would look on the ownership of government securities or a savings account.

Augmenting this trend has been an increase in the number of professional managers who do not have large stock interests in the firm. Although technically these men are given their position and authority by the stockholders, in fact they are generally the absolute authorities subject to control from the management relationships within the firm.

Another consideration in this interface is the phenomenon of mergers. In this case, stock is often used as the medium of exchange, and voting rights once again become important in terms of the control of management. Similarly the phenomenon of a stock raid uniquely ties stockholding to the control of management. In such a situation, proxies are solicited and/or stock quietly acquired until sufficient stock is in hand to vote the current management out and the raiding management in.

Although the relationship of the multitude of small stockholders to the firm involves a myriad of minor systemic impacts which tend to maintain equilibrium, the phenomenon of stock raids and the phenomenon of mergers create major impacts on the firm in a systemic sense. Such events clearly modify existing relationships even if they are not successful and tend to place the system in a state of disequilibrium for a period of time. There is much known concerning relationships with small stockholders but much still to be learned about the dynamics of mergers and stock raids. Since these factors are intertwined, their complexities need to be better understood for effective systems work to take place.

In general, enough is known about normal stockholder relations for the firm to measure environmental relationships in a systemic sense for decision making purposes. The other phenomena, however, require much more observation and study before the questions *what* exists and *why* can be satisfactorily answered.

Among the other objects of the environment, there exists a complex set of relationships concerning the stock of a firm. Its value on the stock market is determined not only by the performance of the firm but also by the performance of its competitors, its industry, its customers and suppliers, and the labor unions with which it is involved. In a broader sense community factors, financial conditions affecting banks, and government actions also affect stock value. Thus, in the environmental set established at the outset in this chapter, all components can have an effect on stock value as they interact as a set of systems.

Some of these relationships have been studied thoroughly and thus predictions can be made of expected results of particular stimuli from environmental objects. Other relationships provide a challenge for further research, however, and so it is not possible at the present state of the art to predict with accuracy what the market value will be for the stock of a given firm at a given time. Certain government statements with implications for increased taxes or war have imprecise effects, technological breakthroughs by competitive firms have imprecise effects, and threatened strikes have imprecise effects to cite but a few examples. Add to this the inexactness of the market psychology as manifested by unusual stockholder behavior and you have the makings of a "stock market game" rather than a stock market science.

Here again, operations research provides insights in defining relationships and measuring attributes. Elaborate correlation studies have revealed certain cause-effect relationships. Indeed, mathematical models have been built whereby the market can be simulated with respect to particular stocks and predictions made of the market impacts on stock value. This type of simulation is now providing some new insights based on observations of a phenomenon and accurate measurement of the consequences of changes in the phenomenon. It is the type of inquiry which will lead to the linkage of this environmental object, stockholders, in a meaningful systemic manner to the operation of the firm.

The Bank-Firm Interface

The basic tie between the firm and the banks with which it deals involves a flow of money and credit to the firm from the bank and a flow of repayments of money from the firm to the banks. In addition, a number of services are provided to the firm by its banks. These include the maintenance of checking accounts, savings accounts, short- and long-term loan arrangements, mortgage services, and the like. If one expands the definition of a bank to that of a financial institution, further services might be noted such as sale-lease-back arrangements, credit analysis, consumer charge plans operated for the firm, consumer loans channeled through the firm but assumed directly by the institution, provision for the acquisition of government securities, stocks, and bonds, and the multitude of new "management services" which range from data processing for the firm, handling tax matters, improving accounting systems, to general management consulting.

Fundamentally, a bank has money for sale. Since the firm requires funds beyond its own capacity to generate them at times, and since a bank also provides a repository for excess funds at points in time, there is clearly a basis for a symbiotic relationship between firms and banks. Neither can survive without the other except in very unusual circumstances. From a systems point of view, the relationship involves a flow of funds from the bank to the firm and back to the bank. This involves a flow pattern whereby the bank and the firm both profit as the funds flow through them. Disequilibrium tends to occur whenever the flow is restricted or stops and when the margin on the use of the funds is reduced to a point where either party is motivated to restrict the flow.

As with the case of the firm-stockholder interface, a substantial amount of data have been gathered to describe what the nature of the relationship appears to be. Elaborate analysis has been made of relationships both within the firm and within banks, since the unit of measure, dollars, is relatively stable and uniform. Governments, principally federal and state, have developed laws concerning banking which require the disclosure of operating data and provide publication of significant ratios on a periodic basis.

In addition, economists and financial analysts have devoted a considerable amount of study to the firm-bank interface since that greatest of all states of economic disequilibrium, the Great Depression, took place. On the basis of the tremendous amount of quantitative data available about bank-firm interfaces and the relatively objective analysis of it during the last thirty years, there is great promise that this object of the environmental set will be well understood in terms of answering the *what* and *why* questions in a systemic sense in the future.

Recently, the bank-firm interface has been subjected to systemic manipulation via the impact of another object in the environmental set, the government. The so-called New Economics is based in part on the assumption that the manipulation of Federal Reserve policy and rates to banks can maintain the economy in a healthy state of dynamic equilibrium. Of course, other measures such as tax policies also are an integral part of this approach. Such an approach is based on systems concepts involving changing a variable with the expectation that the immediate impact on banks will have a subsequent and measurable impact on firms as a second-order systemic response.

Although the structure of the relationships seems fairly reasonable, the magnitude of such changes and their timing require further study. Recent occurrences of system disequilibrium indicate that we still have much to learn about both the magnitude and the timing of such changes.

There are other objects in the environmental set which affect the bank-firm interface in addition to the relationships with government. Banks have relationships with the firm's customers, its suppliers, its competitors, its stockholders, its labor unions, and its community. In these cases there are fewer hard data about the relationships. The current state of knowledge of what exists concerning each of these interfaces as it affects the firm is limited. However the relationships are not insignificant.

In summary, the bank-firm interface is one characterized by a sizeable store of information as to *what* the magnitudes of variables affecting the relationship are. In this sense there is a source of information to answer the question *What* is the state of the relationship? A substantial amount of work has also been directed toward answering the question *Why* do these relationships exist and *why* do certain impulses affect the linkage as they do? Much more remains to be done with respect to the interfaces of the banking institutions and other objects in the environmental set of the firm.

The Supplier-Firm Relationship

The basic relationship between suppliers and the firm is a flow of materials and supplies to the firm from the supplier and a flow of money, and occasionally returned supplies, to the suppliers from the firm. In general, then, the relationship of the firm to its suppliers is similar to the linkage between the firm and its customers. In this case the firm is the customer and the concerns it has for its customers are felt by its suppliers. Implied in the relationship is also the factor of inventories which serve as buffers in the imbalances created by systemic flow patterns characterized by non-continuous demand and/or supply.

As in the case of marketing, most procurement activities are recorded in detail by the firm. Records indicate which parts, materials, and supplies have been procured by the firm in the past, what quantities were purchased, and the prices paid and terms met for their acquisition. This data bank provides much detail for describing *what* exists. The answer to the *why* question is not so clear.

In most discussions of the objectives of the purchasing man's job there are references to getting the right product, at the right price, at the right time. Although this appears to be an admirable objective, no one seems to be able to determine what is the "right" product, price, or time.

The typical procedure involved in the purchasing transaction involves the preparation of a requisition indicating the item needed, the number of units involved, and the time when it is needed. This in turn triggers issuance of the item from the stock of the firm; if that is depleted, then vendors are reviewed and a few selected from whom bids are solicited. Once the supplier quotations have been received they are analyzed in terms of price, discounts, quality of work, service, delivery date, and so forth. After the vendor is selected, a purchase order is made and transmitted to the supplier which starts his cycle of supplying the materials requested. Often the firm employs follow-up procedures to see that the materials are shipped on time. Upon receipt they are inspected, stored, and issued to the requisitioner. Necessary completion of records and payment of the vendor is then accomplished.

This standard procedure exemplifies the way in which firms, in dealing with their suppliers, can suboptimize. Emphasis on price, discounts, and delivery dates has led in some cases to selection of vendors who could not follow through on their commitments.

Emphasis on price and reliance on the nature of the item requisitioned as the trigger impact for the procedure has also minimized the question of substitute materials which may be available. This form of suboptimization has led to the development of the concept of *value analysis*. With this concept the focus is on the function which the material is to serve. Then, on the basis of analysis of this function, that vendor is finally selected who can provide the most value in terms of the need at hand. Such a process requires close working relationships among suppliers and the firm.

A similar new concept is *systems contracting*, which has arisen from another shortcoming of the traditional purchasing procedure. That shortcoming is that the same elaborate procedure tends to be used for almost all inventory items even though some have higher usage than others or represent larger investments. Systems contracting is an approach to streamlining and simplifying the procedures involved in acquiring the majority of low value, low and moderate usage items. With such an approach the relationship with

the vendor is much closer than it is with normal purchasing procedures. Indeed, sole-source purchasing is used, shipments and inven-
torying risks are shifted to the vendor, and billing and payment
procedures are batch processed at periodic intervals rather than as
each transaction takes place.

It is not our purpose to point out some of the procedural problems
in the firm-supplier interface at this point. They do, however, indicate a much larger problem that is concerned with an understanding
of the mutually supportive roles the firm and its suppliers play. By
viewing the supplier as an extension of the firm on the input side
of a system it is possible to examine substantially different relationships than exist today. Such symbiotic relationships can result in
a degree of synergism which substantially transcends that which
currently exists.

Unfortunately, most relationships with suppliers are something
like those with labor unions. Since there is usually more than one
supplier for every item needed, it is possible to play one off against
the other in terms of price, discounts, delivery, and service. The real
question is whether these measures of effectiveness and such fuzzy
ones for the firm as "right" price, "right" quantity, and "right" time
lead to the most effective set of relationships.

In the area of supplier-firm interfaces we have a good deal of data
concerning past practice. There are also some data developing around
newer approaches such as value analysis and systems contracting.
Finding out the answers to the question *what* exists in this area
should not be difficult. The real concern is the question *why*. Some
progress may be made here in the future, particularly with the new
premises underlying such practices as value analysis and systems
contracting.

The relationship of the firm to its suppliers is clearer than the
relationship of the suppliers to other objects in the environment.
For example, suppliers have relationships with the competitive firms
of the firm in question. These relationships can have significant
consequences for the firm. Yet the nature and motives of such relationships are not always clear; indeed, they are often kept secret.

In addition the government plays a substantial role, not only with
respect to legislation affecting trade practices but also as a primary
consumer itself. The massive requirements of the government in
wartime can create severe system disequilibrium for firms in their
dealings with suppliers. In extreme circumstances this can lead
to national rationing of scarce commodities. In lesser circumstances it may lead to the establishment of priorities for firms which
wish to purchase certain items. The determination of *what* exists

under conditions such as these and the development of rational schemes for resource allocation are both difficult and disputable. Much work remains to be done in this area in the future on a broad systemic scale to shed adequate light on the relationship.

Another interesting relationship in the environmental set is that involving suppliers who are also customers of the firm. The consequences of symbiosis become obvious rapidly in such a case, going under the name of reciprocity. In effect, if the firm buys from the supplier, the supplier will buy from the firm. The only trouble with this closed loop is that it occasionally yields suboptimization for both parties. They overlook the broader systemic impacts created by such an alliance.

Another important set of relationships in the environmental set involves the firm, its suppliers, and their labor unions. When one supplier is struck it often disrupts the equilibrium of the system. To overcome this, some firms deal with several suppliers for the same item. With broad scale bargaining, however, even this type of hedge may be marginal. This again is an area in which further study needs to be made to determine the best course of action under systemic impacts.

In the supplier-firm interface many data are available to provide answers to the question *What* exists? Far fewer are available to explain *why* the relationship is as it is. In the environmental set, the relationships among the objects—government, competing firms, customers, and labor unions—as they impact suppliers, are even less clearly understood. Experiments are being made in management to restructure relationships and modify old practices. Yet, by and large, an understanding of the true symbiotic nature of the firm and its suppliers and the degree of synergism which could be achieved using new systemic concepts has barely been explored.

The Government-Firm Interface

The levels of interfaces of the firm and the environmental object government are many. At the local level relationships exist between the firm and its city and county governments. Beyond that are relationships between the firm and its state government. Finally, and perhaps most importantly, there is the relationship of the firm to the federal government.

In general, the government generates certain outputs which affect business and are used by business. The various governments' authority to provide them is not widely challenged. For example, the government provides coin and currency and regulates banking to provide an effective medium of exchange. It provides and maintains roads, ports, and airports and inspects vehicles, ships, and aircraft to facilitate transportation. Fire and police protection are usually provided by government, as well as education, which serves the needs of business. In addition government provides for defense against aggression and for conservation of natural resources. Many other activities of a similar nature are perceived to be the province of government.

Business firms, in most cases, rely on these outputs. Some firms supplement them with their own police and fire protection and educational programs but even in these cases the firm's contribution is typically in addition to the government's contribution rather than in place of it. To a limited extent, the government supplies products to business. To a greater extent, it provides services and information.

The reverse stream of outputs, from the firm to the government, primarily takes the form of products produced for the government, some services, notably research, provided for the government, and the payment of taxes and fees which provide much of the income for the operation of governments at all levels.

If these two streams of activities constituted the relationship between the firm and government, the interface in this area would not be as cloudy and stormy as it is in fact. There are other aspects of the relationship which dramatically affect both business and government and create frictions and conflict resolution problems within the system. For example, businessmen are controlled in their activities by government. They are influenced in their relationships with labor in terms of laws regulating the processes of collective bargaining. They must maintain minimum wages and cannot require workers to work more than an established number of hours per day. They must maintain safe working conditions for their employees and contribute to their welfare when employees are injured under terms of workmen's compensation laws. They must provide for the welfare of retired employees through contributions to the Social Security system. Fair employment practices affect the manner in which recruitment of workers is undertaken. These activities leave little doubt that in the area of the firm-labor-government interface the government plays a significant role in setting constraints. It also provides external stimuli to labor-management activities to settle disputes and thus directly affects decisions in this area.

With respect to customers and the firm, government plays a role which effectively protects the customers from certain business practices construed to be other than in the public interest. Such activities as those directed toward protecting the consumer from adulterated food, unsanitary processing practices, misleading advertising and packaging, untested drugs, and so forth are cases in point. Most activities in these areas involve government testing, grading, and licensing activities, and closure of a firm could result if conditions warrant it. Recent emphasis on automobile safety, truth in lending, and consumer representation in the government indicate that this area of activity is likely to increase rather than decrease.

The government also plays a significant role in the firm-bank-stockholder-government interface. Initially, a firm wishing to incorporate must do so under the incorporation statutes of the state involved. Once securities have been issued, the market for them is tightly controlled by the government. Trading practices are closely monitored and take place within parameters set by government. Banking regulations similarly control both the structure and activities of suppliers of funds to business and repositories of the funds of business. The roles of the Federal Reserve System and Securities and Exchange Commission represent two of the many agencies involved in this area.

With respect to the competitive firms in the environment of the firm, government is heavily involved once again. The government is dedicated to the idea of maintaining viable competition, which seems to wane under conditions of monopoly. Thus anti-trust legislation developed from common law to state acts and then to federal acts such as the Sherman Act, the Clayton Act, and the Federal Trade Commission Act. Antitrust acts, in turn, focused attention on the questions of unfair competition even where monopoly or attempts to monopolize are not present. The Robinson-Patman Act, relating to price discrimination and other discriminatory practices, is an example of this. In general a host of laws and court decisions have come to shape the nature of competition. Whereas the original meaning of free enterprise followed to some degree the Darwinian model of the survival of the fittest, government activities with respect to competition have significantly altered the model. Most recently the role of the government as consumer has provided a powerful governmental lever to alter competitive practices in addition to the traditional lever of legislation.

Finally, the firm-community-government interface presents yet another area requiring conflict resolution. For some time government has curbed the exploitation of land, mineral, and forest resources. More recently conservation activities have been expanded to consider problems of water and air pollution with stringent laws following in their wake. Within the community itself the urban renewal projects and highway development have affected business by physical condemnation of land and buildings, forcing closure and/or relocation of the firm. The role of government with respect to influencing firms to help solve the problems of racial discord in terms of employment is yet to take final shape.

The foregoing comments with respect to the relationship of the firm and government, as well as the firm and other objects in the environmental set as affected by government, represent only a small sampling of the types of relationships which have been established over the years. It is interesting to note that government controls over business in the past few decades have grown to be so extensive and complex that the government must now put controls even upon itself.

System disequilibrium appears within the government-firm interface when controls impede progress, reduce productivity, or foster inefficiency. The slow processes of patent granting, as one case in point, impede progress. The farm subsidy programs tend to reduce productivity. Mountains of paperwork required by the government tend to foster inefficiency. These are just a few cases of system disequilibrium. Dozens of others perhaps come to the reader's mind with minimal efforts at recall.

How one achieves progress, productivity, and efficiency under such conditions is not clear. One approach, taken at intervals in limited areas, has been the employment of businessmen in government to manage a department on a businesslike basis. These efforts have not been particularly successful due in large part to the differences in the objectives of business versus government, as well as differences in their managerial practices.

A significant improvement, however, was registered by Robert McNamara when he instituted systems analysis in the Department of Defense. Faced with a monolithic organization, he attempted to institute effective management planning, coordination, and control through the development of systems. His effort will have a lasting effect on that department. The real question is whether such an approach can be effectively applied in other departments.

In order to implement effective systems analysis in other departments some understanding of what exists is required as a starting point. After that, the logic involved in answering the question Why do relationships exist as they do within the system? must be explored. In certain cases, the nature of the function has paved the way for systems analysis. The Internal Revenue Service efforts at centralization built around computer analysis and retention of massive data banks has forced a systemic approach. In NASA, the completion of integrated and massive space missions has led to the same sort of approach. In the postal system the sheer volume of activity required by growing demand has caused the collapse of old approaches to management of the function. Systems analysis is being explored in this area too.

Our emphasis, however, is on the firm—not the government—as the system. The central question then becomes, Can we utilize systems analysis effectively with respect to the interface between government and the firm? The preceding discussion would indicate that we have available a vast amount of data concerning *what* exists in this interface. We also can track how the relationships have changed over time. The critical problem concerns the limited understanding we have with respect to the resolution of conflict, not only between the firm and the government, but also between the firm and other objects in the environmental set — consumers, competitors, banks, labor unions, stockholders, suppliers, and the community—as affected by the mediation or arbitration role played by the government.

Many of the factors involved in such conflict resolution situations have been studied in detail. Yet before effective systems work can be developed on a broad scale for business decision purposes, more effort must be made to examine these relationships in terms of quantitative data which can be manipulated and analyzed.

With respect to the question *Why* do relationships exist as they do, and change over time? our knowledge is less complete. The firm is faced with imprecise guidelines when it comes to determination of what decisions will be interpreted as monopolizing. It is unsure whether to pursue an attractive merger. It is plunged into uncertainty when legislation is pending or significant political changes are about to take place. When these problems are solved, managers should be able to predict with some accuracy the effect on the government of their decisions. Similarly, government officials should be able to predict with greater certainty the impact of government action on all the parties involved in the system.

Summary of the Environmental Systemic Construct

Systems analysis begins with an understanding of the inputs, transformation processes, and outputs which exist within a system. In addition, the objects in the system should be carefully defined, their attributes measured, and the relationships among them determined.

If one classifies the business firm as the system, then one can visualize an environmental set which contains objects which have interfaces with the firm. In this chapter the environmental objects considered include customers, competitive firms, the community, labor unions, stockholders, banks, suppliers, and government. In each case, a brief view of the environmental object-firm interface reveals that our knowledge with respect to input-transformation process-output flows is limited. Some of the objects are ill-defined, their attributes difficult to measure, and their relationships imprecise. We have seen that the answers to the questions *What* exists? and *Why*? are fairly precise in some cases, such as the bank-firm interface, and very imprecise in others, such as the community-firm interface.

In addition, it is apparent that the environmental objects have significant relationships among themselves. In a sense, the environmental set of business could also be viewed as a set of interacting systems. This area, in itself, provides fertile ground for challenging research.

If the objective of systems analysis is the incorporation of the environmental objects into the operating system, whereby analysis will reveal optimal or near optimal decisions for the firm, then we must be prepared for results which are suboptimal due to the current state of the art in the environmental area. Although it is hoped that, in time, further research will clarify our understanding of *what* exists in the environmental set and *why*, for the present, a more effective thrust can be made within the firm.

In the remainder of the book only a few environmental impacts on the firm will be tracked and only certain dimensions of given outputs affecting environmental objects will be explored. Further development of a systems approach at the environmental level will rest on a considerable amount of research which remains to be undertaken. It is hoped that the concepts and viewpoints expressed in this chapter will provide guidelines for such research.

The impact of the environment on the firm is omnipresent, especially in terms of the firm's decision system. Although tightly constrained systems can be structured at the subsystem level which exhibit all of the characteristics of closed loops and therefore a capacity for self-correction and self-control, such constructs oversimplify the situation.

A business firm as a system is inherently an open system which, to survive and grow, must be adaptive. This adaptive capability essentially reflects the constant barrage of environmental influences which impact the firm. Even though the firm may appear to operate as a closed system operating to achieve specific objectives over a limited time frame, this construct cannot be maintained over extended time frames. The objectives of the firm must be modified from time to time to reflect not only internally generated goals, but also changing relationships within the business environment.

Having considered the environmental set of the firm, we shall now turn our attention to the firm as a system. This represents a further narrowing of perspective as depicted in Figure 1-1. Our first concern will be to review the current controversy centering on traditional or classical theory and the recently developed alternative theoretical constructs. We shall then examine a systemic construct for the firm within which exist the four flow networks described in Chapter 1.

Conceptual Problems

1. It has been suggested that the environmental set of the firm contains the following objects: government, customers, competitive firms, community, labor unions, stockholders, banks, and suppliers. This viewpoint is built around the premise that all of them directly impact the firm and are impacted by the firm. Yet there are second-order impacts when one views an object as a system. For example, the government could be viewed as a system with an environmental set of its own which would include firms as objects of its environment. For effective systems design and systems analysis, what is the value of such a change in perspective? How far should the analysis proceed with respect to second-order and more remote impacts? What criteria should be used to set such limits?

2. With respect to the customer-firm interface, how can one design a closed-loop information system for effective planning and control? Consider the technical problems of tracking consumption stimuli from widely dispersed and numerous customers of products and/or services.

3. How should one firm define its competitors in the firm-competitive firm interface? Will specifications of product lines or services or industry delineations suffice to provide a sufficiently broad perspective?

4. In the community-firm interface, businesses have recently been confronted with a number of responsibilities such as overcoming urban problems, acting on civil rights, and controlling pollution. In many cases the concerned citizenry or government has forced action on the part of the firm. This process is usually quite disruptive to the firm and the community. How can the firm act instead of react in these situations? Should it shoulder such responsibilities in the first place? What problems in addition to those already laid on the corporate doorstep are likely to emerge in the future?

5.   The relationship of management and labor has been characterized by grievances, negotiations, and other manifestations of combative behavior. Since the relationship of labor and management is fundamentally symbiotic, what approaches could be taken to improve the relationship and attitudes of each group toward the other?

6.   In recent years a number of people have built mathematical models of the stock market in order to predict stock values. What environmental factors would you suggest be included in such models? How should coefficients or weights be determined to accurately relate the relative impact of the several factors involved in such a predictive model?

7.   In the bank-firm interface the government impacts the two through Federal Reserve policy by changing reserve requirements and rediscount rates. The macroeconomic consequences of this in recent years have caused it to be used with increased frequency. How can analysts determine the appropriate timing of such changes and the proper changes in magnitude of reserves and rediscount rates?

8.   Systems contracting exhibits a significant step toward building on the symbiotic relationship of the supplier and the firm. What is the potential for expanding this concept along the channel of distribution in both directions, i.e., toward the customers of the firm and from the suppliers of the firm toward their suppliers? What mechanisms could be developed to facilitate such an extension?

9.   Has systems analysis improved operations in the Department of Defense and the National Aeronautics and Space Administration? If so, in what ways could it be applied beneficially in such areas as the management of the postal system, health, education, and welfare activity, agriculture, and so forth?

Organization Theory: Classical, Neoclassical, and Systems

Classical Organization

5

The conceptualization of the firm as a system involves modification of some traditional concepts of organization and management.[1] In order to visualize the role of a systems approach within the firm, certain traditional management concepts will be explored, their deficiencies explained, and an attempt made to show how these deficiencies can be overcome through the use of a systems approach.

Functionalism: The Classical View Versus Systems

A fairly common concept in traditional approaches to organization is *functionalism*. This idea is based on a classification system for the firm composed of such functions as marketing, production, finance, accounting, personnel, research and devel-

---

[1]For examples of the traditional or classical approach to organization theory see the following: L. A. Allen, *Management and Organization* (New York: McGraw-Hill Book Company, 1959); E. F. L. Breech, *Organization* (London: Longmans, Green and Company, 1957); R. C. Davis, *The Fundamentals of Top Management* (New York: Harper & Row, Publishers, 1951); H. Koontz and C. O'Donnell, *Principles of Management* (New York: McGraw-Hill Book Company, 1964); J. D. Mooney and A. C. Reiley, *Onward Industry* (New York: Harper & Row, Publishers, 1931); W. H. Newman, *Administrative Action* (Englewood Cliffs N.J.: Prentice-Hall, Inc., 1963); W. H. Newman and C. E. Summer, Jr., *The Process of Management* (Englewood Cliffs, N.J.: Prentice-Hall, Inc., 1961).

opment, engineering, and so forth. As a classification system, it presumably clarifies the components of a firm and facilitates an understanding of *what* exists. In addition, it is proposed by many that functionalism, which has its roots in specialization of labor, is beneficial in terms of meeting the objectives of the firm.

That functionalism works is not subject to question. For decades business schools have built their curricula around functional departments. Firms, in turn, often have functional organization structures. Indeed, it is likely that the typical progress of a high school student, through his business career to retirement, is within boundaries established by functionalism. This is so deeply rooted that in college a particular student may be classified by his peers as an accounting type or marketing type, to cite two extreme stereotypes.

Similarly, as the student graduates and joins the firm, he may again be classified as a functional type. In professional and trade meetings these types congregate and form social groups which project the behavior characteristic of their function. As they progress in their business careers, their attitudes are further shaped by their functional affiliation until at last they can represent their function in competition with the heads of other functions in the council chambers of top management.

It may seem inappropriate to challenge the concept of functionalism since it is so widely used and since the current organization and management practices of the firm reinforce it. However, a systems approach by its very nature forces such a challenge. In a systems approach to management within the firm, the emphasis is on the discovery of inputs and their allocation as resources to a transformation process. The process, in turn, yields outputs. In some cases, these flows represent the transformation of materials using such inputs as raw materials, supplies, energy, manpower, knowledge, and money into useful outputs such as products and services available to consumers. In other cases, these flows represent information streams which parallel the physical system. In most cases these systemic conceptualizations flow across functional lines.

An interesting case in point concerns the introduction of data processing equipment in organizations. In most firms the computer is originally introduced to handle the clerical procedures involved in such activities as payroll preparation, accounts receivable, accounts payable, and inventory updating. No particular problems are encountered with this approach since it usually involves replication of existing procedures by a computer. These procedures, traditionally compartmentalized in an accounting function within the organization, create few problems beyond that function. The basic rationale for the introduction of the computer is speed, accuracy, and the replacement of clerks.

Data processing would not create problems within the organization if it simply replicated isolated data processing activities within a function. However, once these basic replications have taken place, management becomes aware that the computer has certain capabilities beyond bookkeeping such as the building and maintenance of large data banks, rapid selective information retrieval, and the capacity for analysis of data followed by exception reporting. At this point the rationale for acquisition of more equipment goes beyond the reduction in the number of clerks. It is built around the opportunities to provide more meaningful information to executives for management decisions.

Once the data processing people are given the go-ahead on such applications, they begin to study and make flow diagrams of information systems which flow throughout the firm. In their inquiry they find they need data from many functional areas which will be centrally processed and which will generate reports flowing to several areas including top management and operating personnel. If the effort is successful, many of the boundaries between functions are breached and heretofore "proprietary" information of a functional department is broadly distributed.

The maintenance of such an information system and its embellishment over time tends to integrate the activities of the firm where, in contrast, functionalism tends to separate activities. As more and more managers begin to reap the benefits of integrated understanding of relationships, the boundaries of functionalism begin to fade.

Beyond the argument that functionalism separates and systems analysis integrates lie other arguments favoring a systems approach over functionalism. One is that firms have grown in size in the past few years and have grown rapidly, especially through mergers. The result of this is that firms, which were once reasonably well understood by their managers, now are so complex and large and change so rapidly that no one really understands the whole operation. The impact of change in technology magnifies this problem.

In addition, such firms require managers who can grasp problems in terms of their broad ramifications and make decisions giving full consideration to these ramifications. Our approaches to education for business and managerial progression patterns built around functionalism tend to create a narrower view and create functional specialization and functional allegiances instead.

In summary, functionalism appears to be a classification system which supposedly reflects the firm. So is the systems approach. Which of the two is more realistic as a frame of reference is open to debate. However, breakthroughs in data processing in business toward the concept of management information systems; the complexity, technological change, and growth of firms; and efforts at rational analysis of decision processes using quantitative analysis, indicate that functionalism is a classification system which often impedes our understanding.

Functionalism, Systems, and Inventory Flow

Inventory management provides an example of the deficiencies of functionalism. With respect to the objectives of the firm, inventories should provide buffer stocks to decouple periodic activities so that customers are adequately served and costs of inventory controlled to the end that satisfactory profits result. These objectives can, of course, be refined quantitatively with statements of customer service probabilities, stock-out probabilities, absolute volume targets at each decoupling point, and rate of flow or turnover figures, as well as dollar values reflecting these statistics.

In a typical firm organized on a functional basis, unintended results are often obtained which are not optimal for the firm even though they appear to be optimal for the function itself. In short, functional optimization leads to firm suboptimization. To be specific, consider the major objective of marketing: customer satisfaction leading to continued high levels of sales. Motivations of marketing personnel therefore lead them to press for large stocks of finished inventories to provide rapid service to customers. In addition, they look for large stocks of everything in the product line to preclude "doing business from an empty wagon."

Production personnel have different motivations. They are not interested in maintaining large stocks of finished products. The sooner products leave the shipping dock the better. They are interested, however, in the maintenance of large inventories of raw materials and supplies as well as enough in-process inventory to take up slack in the production process. They want to avoid down-time due to stockouts and produce in long runs to minimize set-ups, which is contrary to the desires of marketing for many small runs to keep a full line of finished products available at any one time.

While those associated with the production function are finding ways to build up inventories of raw materials and those associated with the marketing functions are developing ways to build up finished products inventories, working at cross purposes all the while, the people associated with the finance function appear periodically on the scene. Their objective is to minimize the investment in inventories since this represents tied-up capital. Not knowing the consequences of drastic inventory reductions in terms of marketing or production, they occasionally force such reductions resulting in temporary disequilibrium. Such disequilibrium may be reflected by stockouts, poor customer service, machine down-time, and the like.

To add more confusion, those associated with purchasing are interested in minimizing inventories too. However, they are constantly on the lookout for price reductions on special deals and quantity discounts which reduce the unit cost. In this way, they meet the objectives of purchasing but in turn overload production with excessive inventories of particular items for a time.

Over several years, production, marketing, finance, and purchasing people develop strategies with one another. Games are played to benefit one functional organization at the expense of another. Misleading information with respect to customer demand is generated by marketing; production puts fudge factors in its requirements; finance starts negotiations with unnecessarily tight budget constraints; and purchasing throws up smokescreens about availabilities of supplies. Although each function may optimize its own objectives in this manner and be rewarded accordingly, the results for the firm as a whole are certainly less than optimal.

This situation need not exist. It is possible to view the flow of inventory items throughout the firm in a systemic sense wherein the result is rational balancing of the various objectives of the firm. Current problems often arise simply out of the existence of artificially created functions. Using the same resources, supply inputs, and consumer outputs it is possible to achieve markedly superior results with the systems concept. The details of such an inventory system are developed fully in Part II of the book.

Span of Control: The Classical View Versus Systems

Traditional management theory and most businesses emphasize the importance of *span of control* in organization and management. The basic idea is that a supervisor can plan and control the work of only a limited number of subordinates.[2] The determinants of such

---

[2]See R. C. Davis, *op. cit.*, Chapter 8, pp. 272-280; V. A. Graicunas, "Relationships in Organization," *Papers on the Science of Administration* (New York: Columbia University Press, 1937); H. Koontz and C. O'Donnell, *op. cit.*, Chapter 12, pp. 216-230; W. H. Newman, *op. cit.*, Chapter 15, pp. 253-271; H. A. Simon, *Administrative Behavior* (New York: The Macmillan Company, 1957), pp. 26-28; L. E. Urwick, "The Manager Span of Control," *Harvard Business Review*, May-June, 1956.

a span are not altogether clear but include such factors as the nature of the work, skills of the subordinates and superior, hierarchical level in the organization, leadership characteristics of the superior, and so forth. It is presumed in management literature that a small span of control will yield better results than a large span of control, i.e., the danger is in spreading supervision too thinly over too many subordinates.

If one follows the logic of developing small spans of control, then another problem is presented. A characteristic of traditional organizations is the concept of hierarchy. If a great number of sub-units are created at the lowest level then additional managers must be provided for them. Since span of control usually becomes narrower toward the upper levels of an organization, this means that additional hierarchies must be added to cover the necessary middle managers who in turn must have limited spans of control. The logic of this approach leads to organizations with more and more levels of superior-subordinate relationships as the firm grows in size and complexity. Along with this come the problems of communication and control which must pass through more people (information filters) in the organization.

To facilitate the solution of these problems, standard operating procedures and policies are developed, chains of command established, and formal channels of communication constructed. The net result of all of this is often undue rigidity, excessive red tape, and empire building. In time the organization ceases to be sufficiently adaptive to meet changing current conditions.

In order to overcome these deficiencies, an organization will develop, without careful planning, something called the informal organization with its companions, the informal communication system and key decision makers. These never seem to be committed to paper because they are contrary to traditional organization and management principles and are subject to change. Their redeeming grace is that they are realistic portrayals of *what* exists and are adaptive.

Why should traditional management and organization concepts be challenged? Because they are non-systemic and probably based on inappropriate models. If one were to trace the origins of current models of organization he would find they were borrowed from the experiences, structure, and practices of two institutions: the Catholic church and military organizations. From these two institutions flowed such ideas as hierarchy, span of control, chain of command, line and staff, and superior and subordinate along with notions concerning authority and responsibility.

All of these ideas are not challenged by a systems approach, but such an approach suggests that most of them represent an inappropriate model on which to base a business organization. There may be a better basis which the systems approach could supply.

Unity of Command: The Classical View Versus Systems

The basic principle of *unity of command* suggests that every subordinate have only one superior. This avoids having the subordinate responsible to more than one person and receiving conflicting orders from them. It also establishes a chain of command whereby authority and communications presumably can be tracked through the organization. In addition it indicates the route of advancement normally taken up through the ranks.

This concept is widely used in business and is a direct takeoff from military experience. It may not be particularly effective however in a business organization characterized by rapid change and technological innovation. In such an organization the change agents may be subordinates whose technical specializations are so highly developed that they cannot be given orders effectively by a superior since the superior doesn't know how to go about solving the problems at hand. Where projects must be undertaken, group efforts are expended under the guidance of a project leader or leaders. These projects tend to go through cycles where a group is formed, performs its task, and then disbands in an adaptive fashion. Skills tend to be assembled from throughout the organization to solve the problems rather than being permanently assembled under one superior. Indeed, in some cases committees are formed to pass on the work of the subordinates. Such committees violate the one-man, one-boss rule.

If these kinds of conditions exist, can the unity of command principle be said to cover the needs of modern organizations? If not, perhaps a more flexible approach with resources flowing from project to project would be feasible. Perhaps decisions should be made at various points in the organization as information flows to those points at given times rather than through fixed channels. Perhaps even the idea of hierarchy, consistent with unity of command, could be replaced with information flows and decision points wherein a manager who plays an important role with respect to one type of decision is not even consulted with respect to others. Such ideas are inherent in a systems approach. It is difficult to chart relationships which are dynamic rather than static. But there is little consolation in an organization chart built on traditional concepts if it does not reflect the real-world relationships it purports to represent.

Line-Staff: The Classical View Versus Systems

The concept of *line and staff* in the organization and management of a firm implies that the line officials make the crucial decisions and staff personnel gather facts, analyze them, and advise the line officials with respect to line decisions.[3] Most firms operate on this basis, which is borrowed from military models, yet certain activities have developed which blur the distinction between line and staff. Production planning and control departments, usually a staff group, issue production orders directly, schedule men and machines, and authorize the commitment of other resources to the production process. These types of activities appear to be decisions. Traditional management theory concepts of authority cover this exception with the concept of *functional authority*. This idea, in essence, is that the staff group makes the decisions "in the name of the line official." In fact, the staff group makes the decisions itself, and it may well be that functional authority is merely jargon to cover this type of exception in what otherwise serves to be a useful distinction between line and staff.

The only trouble with this explanation is that more and more staff activities involve decision making as the emphasis on specialized knowledge grows. Engineering groups represent a case in point. Personnel staffs represent another. Public relations departments fit into this category. Research and development groups represent perhaps the fastest growing decision making "staff" group today.

If the line-staff concept of management is an arbitrary dichotomy which has been adopted by business and impedes our understanding of it, as well as its effective functioning, what then can replace it?

From the point of view of systems, strangely enough, the kinds of skills possessed by both line and staff are incorporated into decision points. These decision points have a limited degree of hierarchy. The people involved at these points change as the analytical skills required to solve changing problems at the decision points change. The people at the decision points require the specialized knowledge once thought the province of the staff man to cope with the technical, quantitative, and analytical phases of the decision, yet in addi-

---

[3] See R. C. Davis, *op. cit.*, Chapters 11–13, pp. 369–487; G. G. Fisch, "Line-Staff Is Obsolete," *Harvard Business Review*, September–October, 1961; H. Koontz and C. O'Donnell, *op. cit.*, Chapter 15, pp. 262–295; W. H. Newman, *op. cit.*, Chapter 12, pp. 197–217.

tion they need the experience and judgment to weigh qualitative factors as well, which typically was the province of the line manager.

The obvious question arises, How can one man handle both the technical and general management aspects of all the decisions brought to him? In today's business environment, he can't. The only way this can be accomplished is to focus his attention on certain types of decisions, bring other people's skills to bear jointly in making the decision, and handle such decisions only when the bulk of the relevant data has been processed and analyzed prior to decision time. The decision process itself may involve simulation of the particular subsystem being examined, where the decision group alters one and then another of the variables involved to test results. Deterministic relationships will be explored, probabilistic values utilized in the areas of uncertainty, and then finally the qualitative and subjective elements added to the analysis to arrive at the decision.

If the foregoing observations with respect to functionalism, span of control, unity of command, line-staff relationships, and other aspects of traditional management and organization are valid, then what concepts could apply to the firm which would provide a more realistic model of what exists? How would a systems concept provide a better approach to solving the operational problems of the firm? What type of organization is applicable to the firm when the systems concept is implemented? And most importantly, would such modification meet the challenges created by rapid growth, innovation, increased technology, and the complexities inherent in large organizations?

In general, the traditional or classical approach to organization theory is widely practiced but suffers from a number of limitations. As Scott indicates,

> Paramount among these problems are those stemming from human interactions. ... the interplay of individual personality, informal groups, interorganizational conflict, and the decision making processes in the formal structure appears largely to be neglected by classical organization theory. Additionally, the classical theory overlooks the contributions of the behavioral sciences by failing to incorporate them in its doctrine in any systematic way. In summary, classical organization theory has relevant insights into the nature of organization, but the value of this theory is limited by its narrow concentration on the formal anatomy of organization.[4]

---

[4] W. G. Scott, "Organization Theory: An Overview and an Appraisal," *Journal of the Academy of Management*, April, 1961, p. 10.

This preoccupation with the formal anatomy of organization creates a management environment wherein structural considerations are paramount. Such a preoccupation leads to rigidity in the firm and stifles its adaptive capability. It also generates a viable and adaptive informal organization and informal communication patterns which, unfortunately, operate as sort of an underground. This problem can be overcome by careful analysis using the systems approach to examine the nature of the informal organization and the routes of and reasons for the informal communication patterns. One must exercise care, however, in the establishment of a systemic construct within the firm, to insure the necessary capability for adaptation. Since the systems approach relies heavily on logic and assumptions of rational behavior, it too can become preoccupied with structural matters even though they differ from those developed in classical organization theory. This is especially evident where communication patterns have imbedded in them formal computer programs which interface with decision makers. Such programs, which often are difficult and expensive to modify, may build a sort of structural rigidity in information systems and decision systems which curtails the adaptivity of the firm.[5]

Before we examine the firm as a system, one other development in organization theory should be explored, the so-called neoclassical theory. Its focus is on the human aspects of organization: the into incorporating the human element in classical organization theory. But although it solves some of these problems, neoclassical theory too has certain limitations, which can be overcome by using the systems approach.

Neoclassical Organization Theory

The neoclassical theory of organizations developed as a reaction to the structural, mechanistic assumptions of classical organization

---

[5] For further consideration of the problems of incorporating a management information system in a firm see the following: R. E. Adamson, *Implementing and Evaluating Information Processing Systems*, System Development Corporation Report, SP-1294, November, 1963; J. I. Barnett, "How to Install a Management Information and Control System," *Systems and Procedures Journal*, September-October 1966; J. Deardon, "How to Organize Information Systems," *Harvard Business Review*, April, 1965; I. S. Gottfried, "Presenting Complex Systems to Management," *Administrative Management*, August, 1965; J. Hockman, "Specifications for an Integrated Management Information System," *Systems and Procedures Journal*, January-February, 1963; I. R. Hoos, "When the Computer Takes Over the Office," *Harvard Business Review*, July-August, 1960; F. Kaufman, "Data Systems That Cross Company Boundaries," *Harvard Business Review*, January-February, 1966.

theory. Its focus is on the human aspects of organization: the informal group, individual interaction patterns, and the dynamics of the informal organization.

The emphasis on the human aspects of organization resulted from the pioneering work of Roethlisberger and Dickson in their Hawthorne experiments.[6] While establishing their research methodology on classical, mechanistic assumptions they found that the results of their experiments did not conform to changes in mechanistic variables but rather to human factors within the organization. From these studies developed the human relations school of management which concerned itself with human factors.[7]

Functionalism: The Neoclassical View Versus Systems

The concepts discussed earlier in the chapter with respect to classical organization theory were modified by the neoclassicists rather than abandoned. For example, the concept of functionalism and the specialization of labor with which it is associated were modified to take into account the group sentiments associated with these functions. Problems of intergroup conflict were explored and means of overcoming these problems were proposed based on ideas of group influence patterns and the authority positions of individual members of the group. The concept of authority itself was modified.

The classical approach to authority suggests that it is given to the person occupying a particular position by top management when the authority associated with that position is established. The neoclassicists would suggest, on the other hand, that authority is earned by a person from his peers and subordinates and that authority cannot be assigned arbitrarily by top management to a particular position

---

[6]F. J. Roethlisberger and W. J. Dickson, *Management and the Worker* (Cambridge, Mass.: Harvard University Press, 1939. For another view of the pioneering phase of the neoclassical school see E. Mayo, *The Human Problems of an Industrial Civilization* (Cambridge, Mass.: Harvard University Press, 1946).

[7]Representative books of the neoclassical school are as follows: Chris Argyris, *Interpersonal Competence and Organizational Effectiveness* (Homewood, Ill.: Richard D. Irwin, Inc., 1962); E. Wight Bakke, *Bonds of Organization, An Appraisal of Corporate Human Relations* (New York: Harper & Row, Publishers, 1950); Keith Davis, *Human Relations in Business* (New York: McGraw-Hill Book Company, 1957); Mason Haire, *Psychology in Management* (2nd ed.; New York: McGraw-Hill Book Company, 1964); Frederick Herzberg, *et al.*, *The Motivation to Work* (2nd ed.; New York: John Wiley & Sons, Inc., 1959); Robert L. Kahn, D. Wolfe, *et al.*, *Organizational Stress: Studies In Role Conflict and Ambiguity* (New York: John Wiley & Sons, Inc., 1964); Douglas McGregor, *The Human Side of Enterprise* (New York: McGraw-Hill Book Company, 1960), and *The Professional Manager* (New York: McGraw-Hill Book Company, 1967); Victor H. Vroom, *Work and Motivation* (New York: John Wiley & Sons, Inc., 1964).

and yield effective results. Although such a modification to classical theory represents a more realistic insight into the dynamics of organization, it does not address itself to the more fundamental question: Should functionalism be replaced with an alternative construct which would serve inherently to integrate activities rather than separate them and thereby reduce the fundamental cause of intergroup conflict? This question is raised by those taking a systems approach. The systems approach provides a significant alternative to classical theory rather than the modification of interpretation proposed by the neoclassicists who fundamentally accept the classical structure as a given.

Span of Control: The Neoclassical View Versus Systems

The classical approach to the span of control concept presents a structural dilemma. If the span of control is small the result is supposed to be improved supervision, but the trade-off is that a tall organization with many hierarchical levels will result. The problems associated with this are an excessive number of supervisory positions, red tape, cumbersome communication channels, and so forth. The establishment of large spans of control overcomes many problems associated with multiple hierarchical levels; yet, as a trade-off, it minimizes the effectiveness of supervision. Such a view presumes that there exists some discrete numerical relationship of supervisors to subordinates. In approaching this dilemma some classicists have concerned themselves with determining the appropriate span of control in terms of absolute numbers or universally applicable ratios.

Scott presents his evaluation of the neoclassical reaction to the classical position in this way.

> An executive's *span of control* is a function of human determinants, and the reduction of span to a precise, universally applicable ratio is silly, according to the neoclassicists. Some of the determinants of span are individual differences in managerial abilities, the type of people and functions supervised, and the extent of communication effectiveness.
>
> Coupled with the span of control question are the human implications of the type of structure which emerges. That is, is a tall structure with a short span or a flat structure with a wide span more conducive to good human relations and high morale? The answer is situational. Short span results in tight supervision; wide span requires a good deal of delegation with looser controls. Because of individual and organizational differences, sometimes one is better than the other. There is a tendency to favor the looser

form of organization, however, for the reason that tall structures breed autocratic leadership, which is often pointed out as a cause of low morale.[8]

Although the neoclassicists chide the classicists for their ratios, the neoclassical guides to appropriate spans of control are not much more helpful in terms of the problem of designing an organizational structure. These guidelines presume one knows about the personality of particular individuals in determining the proper span of control. Attributes such as managerial ability, communications effectiveness, leadership style, and effectiveness in human relations are among those one should consider, according to the neoclassicists. If this is true, as it seems reasonable to assume, then the question arises: What happens when transfers and promotions take place in an organization which has been staffed according to these guidelines? It would seem to follow that spans would have to be continually modified according to the particular array of skills particular managers may bring to their new jobs. Such frequent expansions and contractions of groups would, of course, adversely affect the informal group cohesiveness and group identity which the neoclassicists propose be maintained.

Such examples of how "good human relations" applied in one area at one point in time creates problems in another area at another point in time are numerous. This is an unfortunate dimension of neoclassical theory. Although a good deal of thought has been given by neoclassical theorists to the solutions of individual problems in the industrial setting, the neoclassical approach fails to provide integration and balance in a systemic sense.

As with the case of functionalism, the neoclassicists have accepted a classical concept, span of control, and attempted to modify it in light of human relations insights. With the systems approach, the relevancy of such modification is questioned. The fundamental question concerns whether span of control is a viable and necessary concept in the design of a system. If the system design does not depend on a hierarchical structure and military models of superior-subordinate relationships, then there is the possibility that the concept can be abandoned. In its place might appear similar questions but they would concern the number of input information channels which should impact a decision center, the nature of the information, the complexity of the decision processes carried out at that decision center, and so forth.

---

[8] W. G. Scott, *op. cit.*, p. 12.

Unity of Command: The Neoclassical View Versus Systems

The classical position with respect to unity of command is that no subordinate should have more than one superior. Its derivation is from military models in which orders are given by a superior and followed by a subordinate. In this manner, a chain of command can be established. The concept assumes a hierarchical structure in which authority is vested in positions rather than in individuals.

As the neoclassical school developed, it observed, quite correctly, that a business firm is not a military organization and that leadership in industry is not a function of position but a characteristic of a man. Many studies of the human relations group have focused on what the characteristics of effective leaders are and, as might be expected, the attributes discovered are many, varied, generally too virtuous to cope with human frailty, and occasionally contradictory. Some people propose that an autocratic style of leadership is best, others favor a democratic style, while still others suggest that benevolent autocracy is the answer.

Like most neoclassical theory, the concepts associated with the unity of command concept reflect a preoccupation with the human side of enterprise. In general, these concepts accept the classical assumptions of hierarchy and superior-subordinate relationships and tend to build on them. A notable exception is Douglas McGregor's work which introduced the Theory X versus Theory Y controversy.[9] This contribution was very significant since it challenged some of the basic assumptions about motivation and leadership patterns. What is needed, if the neoclassical school is to regain its former influence, is more theoretical development of this type — the type of research which challenges existing premises and principles.

The systems approach presents such a challenge. Although no unified theory has yet emerged, there is promise in approaches such as matrix management. This concept is a distinct alternative to the classical concept of unity of command. Under this scheme, the unity of command concept is abandoned and replaced with a situation wherein an individual may work with several "superiors" rather than just one. On one axis of such a matrix may be the managers of several projects. On another axis may be the managers of a number of resources. Such a situation is depicted in Figure 5-1.

In such a matrix as that depicted in Figure 5-1, the Project A Manager is responsible for the completion of his project and he

---

[9]D. McGregor, *The Human Side of Enterprise* (New York: McGraw-Hill Book Company, 1960).

	Materials Manager	Manpower Manager	Money Manager	Machines, Facilities, & Energy Manager
Project A Manager				
Project B Manager				
Project C Manager				
Project D Manager				

Figure 5-1.   Matrix Management

draws upon the resources which are managed by the Materials Manager, Manpower Manager, Money Manager, and Machines, Facilities, and Energy Manager. These managers reflect concern over the maintenance of the four flow networks mentioned earlier in the book and developed more fully in the next chapter. The Project B, C, and D Managers, of course, also draw upon the pool of resources as the requirements of their projects dictate.

On the other dimension of the matrix are the resource managers. Their responsibility is the management of particular resources and they draw upon the project managers to determine how they can manage effectively these resources in the interest of completion of the projects within time, cost, and performance constraints.

Neither group in matrix management is "superior" nor "subordinate" to the other. There is Y axis liaison among the project managers and X axis liaison among the resources managers to provide integration and balance. A chain of command does not exist at this level. The question therefore may be raised, But doesn't this create conflict in the organization? The answer is emphatically yes. As a matter of fact a large part of the management process under this concept involves conflict resolution. The important point is that the conflict resolution process is out in the open rather than suppressed within the organization. And this process is in the interest of integration of the activities of the firm rather than of solving specific conflicting problems without reference to their impacts throughout the organization.

As an example of this, consider the materials management function. Where modifications in material allocation patterns are made, they are viewed as they impact all of the projects simultaneously. Or, looking at the matrix from the other axis, if the manager of

Project *A* encounters difficulties in performing his task, he consults all of the resource managers who *balance* the reallocation of resources of all types to assist him in overcoming his problem. These readjustments, of course, are made known to the other project managers if their projects are impacted by the change.

Although there are several such innovative concepts being developed, this one serves to provide an example of how a systems approach, which stresses *integration* and *balance,* effectively challenges such existing premises as unity of command. Even the term *manager* takes on a different meaning. The "manager" of the materials flow network may be a number of people who work together as an integrated team and fulfill managerial responsibilities as a group rather than as individuals.[10]

Line-Staff: The Neoclassical View Versus Systems

Another example of neoclassical modifications to classical concepts concerns line-staff relationships. In classical theory, line officers carry position authority and make the decisions; staff personnel exist to provide advice in their areas of expertise. The neoclassicists recognize that this distinction is inherently difficult to maintain, especially in light of the increased complexity of management and the explosion in knowledge characteristic of the rapid development of technology.[11]

The human conflicts which result from staff usurping line authority to implement their ideas and the counter move by line managers to ignore and isolate staff groups from the action, provide numerous case studies in human relations which the neoclassicists use to derive methods of resolving the conflict. Resulting from such studies come such admonitions as the following:

— Staff personnel ought to have clearly defined authority so

---

[10]For further insights into the concept of matrix management and the environment within which it is being used see the following: D. Cleland, "Understanding Project Authority Requires Study of Its Environment," *Aerospace Management,* Spring-Summer, 1967; P. O. Gaddis, "The Project Manager," *Harvard Business Review,* May-June, 1959; J. F. Mee, "Matrix Organization," *Business Horizons,* Summer, 1964; C. Middleton, "How to Set Up a Project Organization," *Harvard Business Review,* March-April, 1967; J. M. Stewart, "Guides to Effective Project Management," *Management Review,* January, 1966.

[11]See the following for additional views on line-staff: M. Dalton, "Conflicts Between Staff and Line Managerial Officers," *American Sociological Review,* June, 1950; G. G. Fisch, "Line-Staff Is Obsolete," *Harvard Business Review,* September-October, 1961; J. M. Juran "Improving the Relationships Between Staff and Line," *Personnel,* May, 1956.

they know where their authority ends and line authority begins.

— Line personnel should utilize effectively the talent represented by their staff units and respect their specialized knowledge.

— Staff personnel should have clearly stated objectives so that their activities contribute to the fulfillment of organizational objectives as well as to the fulfillment of personal needs.

— Staff and line personnel should engage in participative planning and other activities which will provide better communication.

Although these kinds of propositions reflect a sensitivity to human emotions in an industrial setting, they fail to raise the fundamental question: Is the basic line-staff concept really useful in the organization or does it represent an industrial application of a military model which is fundamentally inappropriate in a business organization?

The systems approach suggests that the way to reduce line-staff conflict is not to enforce line-staff distinctions as the classicists would propose or to develop conflict-resolving admonitions as the neoclassicists would propose, but rather that the fundamental line-staff concept be abandoned. In its place would exist an entirely different organizational pattern which combines the skills of line and staff personnel in decision makers who are components of decision systems interfaced with information systems which track physical flow networks.

In essence, the system proponent would not accept a neoclassicist modification of a classical structural proposition which is faulty in the first place. Simply patching up a fundamental problem with "better human relations" does not solve the problem. The dictum most often encountered in this respect is the often repeated and shallow comment of human relations case analysts: "This whole mess just boils down to a communications problem—what we need around here is better communications!" When pressed to explain what he means by "better communications," the neoclassicist often replies "*something* which would improve understanding among these people — they just don't understand one another." At this point a proponent of the systems approach stops pressing the neoclassicist since the prospect of his developing an integrated management information system as the "*something*" is dim indeed.

Summary of Organization Theory

In this chapter, four fundamental concepts associated with management and organization are discussed: functionalism, span of control, unity of command, and line-staff relationships. In each case the classical view is explored and some deficiencies are noted. The concern for the formal anatomy of organization exhibited by the classicists results in a number of problems which may be overcome through the use of a systems approach to organizations.

The neoclassical or human relations school is also examined as it bears on functionalism, span of control, unity of command, and line-staff relationships. In each case, the deficiencies in classical theory are partially overcome by the neoclassicists through their study of interpersonal relationships, conflict resolution, and so forth. Although such efforts tend to mask the underlying problems through temporary alleviation of symptoms, they do not solve the fundamental problems facing managers today. In order to do so, the basic premises of organization, such as functionalism, span of control, unity of command, and line-staff relationships, must be subjected to close scrutiny. The systems concept not only subjects them to scrutiny but also suggests that many of them may not be appropriate premises for management.

For example the thrust of functionalism and specialization of labor is separation and differentiation. The thrust of the systems approach is just the opposite—integration and the effective linking of activities to meet the objectives of the firm. Such integration often requires a significant modification of current views of functionalism.

Although this chapter cites four fundamental concepts of organization and presents the classical, neoclassical, and systems view of them, it merely scratchés the surface of the controversy in organization theory.[12] Many schools of thought today address the problems of organization and management. Ultimately, the contributions of all of these schools of thought will provide a better understanding of management. Some of them may be merged as they become more fully developed. At any rate, the controversy is a healthy one. It is providing fresh insights and many new perspectives. The systems approach is one of these. It holds much promise as a viewpoint. In

---

[12]For discussions of the controversy in organization theory, see the following: H. Koontz, "The Management Theory Jungle," *Journal of the Academy of Management,* Vol. 4, No. 3, December, 1961, and *Toward a Unified Theory of Management* (New York: McGraw-Hill Book Company, 1964); J. G. Hutchinson, *Organizations: Theory and Classical Concepts* (New York: Holt, Rinehart & Winston, Inc., 1967); E. P. Learned and A. T. Sproat, *Organization Theory and Policy* (Homewood, Ill.: Richard D. Irwin, Inc., 1966); L. Sayles, *Managerial Behavior* (New York: McGraw-Hill Book Company, 1964).

the next chapter we shall examine the firm as a system. Such an examination will provide a more specific exposition on how the problems cited in this chapter can be overcome through the use of a systems approach.

Conceptual Problems

1. Functionalism is a common practice in most firms; yet in that it enforces differentiation of tasks, it becomes more difficult to achieve effective integration of tasks. The systems approach has been proposed as a concept which may effect such integration, yet this may require abandonment of traditional functionalism. What alternative organizational concept might provide such integration without the abandonment of functionalism?

2. Functional optimization leads to firm suboptimization. Examine your experiences in organizations and comment on the validity of this statement.

3. Administrative and organizational concepts such as hierarchy, span of control, chain of command, line and staff, and superior and subordinate relationships were borrowed from institutional models; namely, the Catholic Church and military organizations as the theoretical underpinnings of traditional organization theory. They are now broadly used in business. How can alternative approaches, including the systems concept, be introduced in business? How would you counter the argument that the present concepts work and the newer ones may not work?

4. What do you see as the organizational consequences of incorporating the systems concept in a firm? Consider the short-run consequences in terms of conflict resulting from such a change as well as the long-range impacts on modified organizational relationships.

5. The unity of command concept may be violated repeatedly where the systems concept is used in organizations. A matrix concept, found in the management of research projects, is a case in point. What are the ramifications of this in the process of the resolution of conflict in the organization?

6. The neoclassical or human relations school places heavy emphasis on interpersonal relationships within the firm. Although interesting insights have been developed by this school, there is some question as to whether any universal truths have emerged. Can you identify any such universal truths?

7. What are the determinants of business leadership? How would you utilize your knowledge of these leadership traits to develop more effective operations? Does your list of traits favor autocracy or democracy as a leadership style or does the choice of the leadership style depend on the situation in which it is applied?

The Firm as a System

In the preceding chapters we have examined the systems concept and the environmental systemic construct within which the firm operates. The focus in this chapter narrows from the view of the business environment to the view of the firm as a system. This represents the third level in the systemic hierarchy depicted in Figure 1-1 (p. 6).

In this chapter the firm will be examined in terms of several flow networks or systems. The requirements for systems planning and control will be considered as well as some of the managerial and organizational consequences of a systems orientation to the firm.

The Objects in the System

If one were to answer the question *What* exists? when examining a firm without using the classification system of traditional management he could start at a very fundamental level. He would observe that the firm has associated with it a number of people with particular skills and knowledge, machines of many types from typewriters to turret lathes to trucks, money, materials including raw materials, supplies, parts, goods in process and finished products, facilities such as buildings, and energy sources such as gas and electricity.

In its static state a firm is just that—a collection of resources. Yet a catalogue of all the items in the firm, an inventory of its resources, reveals little about its nature as an organism. This level of inquiry is equivalent to the medical profession's knowledge based only on the observation of objects: arms, legs, eyes, organs, and so forth.

It is only when the firm is in a dynamic state that managerial activities have meaning. In its dynamic state a firm could be viewed as an organism composed of systems which transform the resources into useful products and services. Then the question arises, What are the key objects of such a system? What is the role of management within the system?

It would appear that one function of management is the planning of the allocation of resources and their transformation processes. The complementary management function associated with planning is controlling. The latter implies the former if control is defined as seeing that the results being achieved by the system conform to the plans. If it is useful to consider planning and controlling as the key functions of management, the next question concerns the focus of planning and control within the firm. What shall be planned and what shall be controlled?

The Materials Flow Network

In reviewing the listing of items found to be associated with a firm, it seems reasonable that one of the key objects in the system is the product or service being produced. In this sense one could view a flow pattern or system through which materials move. Without going into the environmental set objects in any depth, one might depict as inputs to this system the materials and supplies provided by the outside suppliers. Plans would have to cover the acquisition of such materials and supplies, their transportation to the plant, their storage, allocation, and transformation within the plant, their transportation and storage enroute to consumers, and final sale to the consumer. A system built around such a flow network would effectively link suppliers and consumers in the environmental set by the transformation processes of the firm with respect to the planning and control of materials flow.

Such an integrative approach incorporates into one system a number of activities which in traditional functionalism would have fallen into the departments of production, marketing, purchasing, and traffic management. Under such a functional approach, stimuli resulting from retail sales usually trigger delayed replenishment from wholesalers who, in turn, trigger delayed replenishment from production which, in turn, through more delays, triggers requisitions for replenishment of stocks at the plant from the suppliers, handled by purchasing. As has been noted earlier, if handled separately, these functional units will tend to optimize their individual objectives instead of the firm's.

In addition is the problem of oscillation and magnification of fluctuations which is a function of delays and misleading information.

By using a systemic construct, in which retail sales trigger responses throughout an integrated system all the way back through the system to suppliers without significant delay, such oscillation and magnification could be substantially reduced. Further, the objectives of the firm would be primary rather than secondary, without an artificial division of functions along the material flow continuum.

The Money Flow Network

A second flow system which would be useful to plan, track, and control concerns money. In its function as a medium of exchange, its flow pattern would be generally toward the firm from banks, stockholders, and customers in the environmental set and away from the firm toward government, labor, suppliers, and the community in the environmental set. In its function within the firm as a measure of resource allocation, money flow patterns could be established to balance the trade-offs requiring the homogeneous quality of money as a measure of dissimilar attributes. Using this flow network as a focal point for planning and control, one would concentrate on maintaining the proper rate of flow and volume controls over time to balance both the input and output streams of funds to and from the environment. Within the firm, resource allocation decisions would be based primarily on the balance of rate of flow and volume over time of the physical system, with money only playing a role as a unit of measure. Other measures such as physical measures of materials and capacities would provide the focal point for analysis, rather than dollars. The distinction is significant.

Money as it circulates to and from the environment with respect to the firm is a commodity, i.e., it is tangible and leaves one system to enter another. Money as it circulates within the firm is only a unit of measure which, in many cases, clouds or colors our understanding of real phenomena because it is not particularly characteristic of particular objects within the system as are other measures such as weight, units, or dimensions. In the past, most emphasis has been placed on external money flow patterns through the activities of finance with respect to banks, stockholders, and government; purchasing with respect to suppliers; marketing with respect to customers; and accounting with respect to labor. Public relations, more than any other organization, responded to outward money flows to the community. Such a hodgepodge of separate responsibilities with respect to environmental money flows results in many cases of imbalances requiring emergency borrowing from banks or heavy nonproductive cash reserves. By focusing on environmental cash flow in its entirety and integrating the functions mentioned above toward this end, an improved money flow pattern should result and emergencies should be averted.

The Manpower Flow Network

A third object in the system requiring planning and control is people. Provision should be made for the external flow between the firm and the environment. The responsibilities envisioned here go beyond the personnel recruitment and placement activities on the input side and exit interviews on the output side, responsibilities carried out for years by personnel departments. The manpower flow network also involves the flow of personnel between the firm and other environmental objects, notably the government and the community, both of which in the future may require personnel from the firm or supply personnel to the firm.

Internally, the critical aspects of manpower flow are concerned with the allocation of the manpower resources in various areas for the immediate planning period ahead as well as manpower flow considerations associated with motivation, training, and promotion of personnel. Under the traditional functional approach it is common to find that the functional managers carry the authority to assign and promote their personnel. The personnel department is concerned with training. Everyone, yet no one, works full time at motivation.

It is difficult to conceive of another scheme of internal manpower flow planning and control since we are all inheritors of hierarchical and departmentalized environments. However, such an approach may solve many problems. For one thing, an approach might be found that would treat every man as a man rather than a type. It would be common to find a man starting in production activities, moving to sales, financial, and personnel activities, even in and out of engineering and accounting as his skills allow, to emerge a prime candidate for general management rather than being boxed into a function as is now the case. It should minimize empire building since manpower planning and control is beyond the individual scope of operational supervisors.

The Machine, Facility, and Energy Flow Network

A fourth area for planning and control concerns the allocation of resources in terms of machines, facilities, and energy. The responsibility here involves both acquisition of materials from suppliers in the environment and internal allocation of resources within the firm. Currently within the firm such resource allocation decisions are made by each functional department with respect to its facilities with periodic battles among departments when they want the same facility. In finance such problems are moderate. In marketing and office management they become more severe. In production, machine and facilities allocation becomes a daily critical balancing act.

By placing the focus on energy sources in terms of volume and rate of flow and evaluating machines and facilities in terms of capacities and balance with respect to manpower and materials flow patterns through them, it should be possible to achieve a more balanced operation. Under this approach the acquisition of new materials and facilities would not take place in negotiation between the requisitioning agent and accounting or finance. Rather, they would occur periodically when all such requests affecting capacity and the effectiveness and efficiency of the operation are weighed simultaneously and approvals given, which tend to fulfill the objectives of the firm rather than simply the objectives of a given department. There is a sort of empire building with respect to equipment which parallels that associated with people. Such a centralized planning and control effort directed toward machines, facilities, and energy should assist in overcoming this.

The Information System

In order for the four systems which we have just discussed to function effectively it would be necessary to develop a rather elaborate information system. This in itself would be a significant flow system within the firm. Its function would be to establish the appropriate sensors and units of measure for data acquisition required by the four systems. It would maintain extensive data banks relative to the decision processes involved in the four systems. When required, it would prepare output information reflecting analysis of data gathered. If at any time exception reports were generated in the daily operations, these would be transmitted immediately without regard to the periodicity of the regular reporting system. And its most difficult responsibility would be to provide information regarding any one of the four systems to the decision points of the other systems in a form where actions taken in one system could be reflected as they impact the others.

In this sense the information system serves the function of integration among the other four systems and is the nerve center for the entire firm. A system such as this would require full time maintenance to keep it responsive to changing needs, for it is in this area in particular that the hazard of rigidity in the system can creep in, especially in terms of reluctance to re-program to manipulate or report data in modified ways. In addition it is important that such an information system be monitored or audited to determine that it reflects what is really happening within the firm. The four basic systems are designed to plan and control real phenomena: materials, money, manpower, and machines, facilities, and energy. The information system is an artificial system which, through symbol manipulation, reflects the real world but is not the real world itself. Therefore, as decision makers rely on the symbols rather than on observation of real phenomena, there must be measures to control the information system so that its symbolic representation of real activities and events does not vary significantly from the real activities and events themselves. A symbolic representation of the concepts presented above is shown in Figure 6-1.

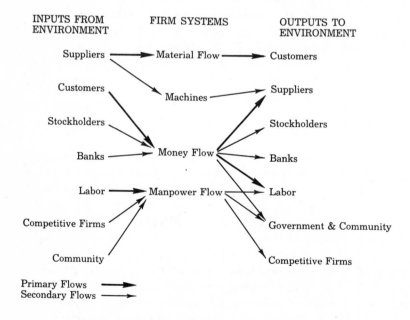

Figure 6-1.   Environment-Firm Flow Network

The Order Flow Network

One other symbolic flow network, the flow of orders, is also involved in the firm. Although it does not represent a flow pattern of real objects, it does serve as a classification scheme. As resources are allocated through management decisions, they are identified by order numbers and order descriptions. As they flow from the input phase through the transformation process into the output phase the resources can be tracked by these order numbers.

Orders do flow through the firm. However, they should not be mistaken for physical objects. They are merely symbolic representations of commitments of resources and serve only as identifiers. In some cases, several orders may be grouped together in those phases of system operation where divergence and convergence patterns are present. For example, several orders may be classified as one during manufacturing operations to effect long production runs. They may also be grouped during transportation processes. It is only when they reach the ultimate consumer that they must be treated as discrete identifiers. And even then, in cases such as open-ended orders which are filled on a continuing basis, the orders may be grouped. Arbitrary grouping which impedes the smooth flow of the physical flow processes must be reviewed carefully to avoid suboptimization of the operations flow processes.

The Key Planning and Control Variables

When utilizing flow networks, three variables are important for planning and control. The first of these is the *capacity* of the system. The second is the *volume* in the system. The third is the *rate of flow*. To visualize the interaction of these variables, consider the mechanical analog in Figure 6-2.

In this physical system, composed of a horizontal pipe with a single standpipe, assume that a fluid is introduced at the input end.

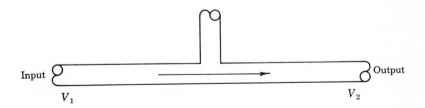

Figure 6-2.   Mechanical Analog of a Flow System

As long as pressure is applied, the fluid will flow through the horizontal pipe and exit at the output end. Such a structure has no capacity for control. As pressure increases on the input end, the rate of flow increases on the output end. When pressure drops to zero on the input end, the rate of flow drops to zero on the output end. Between input and output no capacity for the maintenance of equilibrium is present.

As a first step toward the maintenance of equilibrium, consider the incorporation of valves at points V_1 and V_2. With these installed, we can predict the operation of the system, yet we still fail to maintain equilibrium in the system. For example, if the horizontal pipe is full of fluid and valves V_1 and V_2 are closed in the static state, the volume in the system must be equal to or less than its capacity. When valve V_2 opens suddenly, the system will drain to an empty condition. If, on the other hand, starting from the static state above, valve V_1 opens, assuming back pressure behind it, and valve V_2 remains closed, the standpipe will overflow.

To maintain equilibrium, it would appear desirable to have a feedback mechanism from V_2 to V_1 and a feed forward mechanism from V_1 to V_2. Such a system is depicted in Figure 6-3.

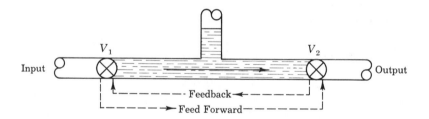

Figure 6-3.   Mechanical Analog of a Flow System
with Feedback and Feed Forward

Considering the system in Figure 6-3, if a given volume is in the pipeline which is equal to or less than the capacity of the system, impacts on the input and output ends can now be absorbed providing the feedback and feed forward information loops operate rapidly. For example, should valve V_2 open slightly, feedback information would be transferred to valve V_1 to effect a similar opening, thus equalizing the rate of flow and maintaining equilibrium. This assumes, of course, a back pressure of available fluid at the input end of the system.

If, on the other hand, valve V_1 were to open slightly, a feed forward signal would be sent to valve V_2 to open the same amount.

Again the rapid adjustment of V_2 to the impact at V_1 would maintain equilibrium.

In this example, it can be seen that volume, capacity, and rate of flow are key elements for planning and control. Effective operation of a flow system requires first of all a clear understanding of its capacity. This will aid in avoiding system overloads. Secondly, the volume in the system is a critical information item as it is related to the magnitude of shock or impact the system can absorb before corrective action must be applied or the system capacity is exceeded. It also is critical if the volume is very low where an impact may void the system before corrective action can be taken.

In addition, to maintain equilibrium the system requires rapid feedback and feed forward among the control points. The longer the delay between the time a signal is sent and received between two points, the greater the risk of delay-induced disequilibrium.

The same principles which apply to the mechanical analog apply to the flow networks of the firm. Materials, money, manpower, machines, facilities, and energy all exhibit flow characteristics and are subject to imbalances created by lack of careful consideration of capacity, volume, and rate of flow — the key variables for planning and control.

Convergence-Divergence Patterns

Another critical characteristic of the planning and control of flow systems concerns their structure as networks. In the simple system described above, the network is minimal. In the firm as a system, however, patterns of convergence and divergence exist. In order to plan and control effectively their operation, it is necessary to track these networks. Such a network is depicted in Figure 6-4.

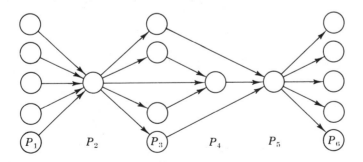

Figure 6-4.   Convergence-Divergence Patterns

In Figure 6-4 a convergence pattern exists from point P_1 to P_2. This is followed by divergence from P_2 to P_3. Partial convergence occurs from P_3 to P_4. Total system convergence occurs from P_3 and P_4 to P_5 followed again by divergence from P_5 to P_6.

Such patterns add another dimension to the problems of systems design and systems analysis. Knowledge of the convergence-divergence patterns as well as knowledge of capacities, volumes, and rates of flow within the network are required for the operation of even modestly complex systems ... and the firm as a system is a complex network by any definition.

As an example of the critical nature of convergence-divergence patterns in the firm, consider the material flow network. In this case, careful consideration must be given to convergence patterns from vendors to receiving at the firm. This must be followed by tracking of the multiple convergence-divergence patterns taken by materials as they go through manufacturing. Finally, an elaborate divergence pattern exists to move materials from shipping at the firm through channels of distribution to a variety of customers.

Similar patterns exist in the other flow networks. Money, men, and energy are routed over a variety of convergence-divergence routes over time as the firm operates. Convergence-divergence considerations also are critical with respect to the allocation of machines and facilities.

Planning and Control of Flow Processes

To develop a managerial model for the planning and control of such flow processes as those associated with materials, money, manpower, machines, facilities, and energy, it is useful first to review some basic concepts of planning and control as they apply to physical systems. Figure 6-5 will serve as a point of reference.

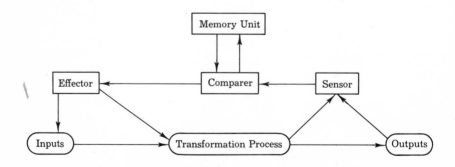

Figure 6-5.　　The Feedback Loop

In this schematic it is assumed that the physical inputs flow through a transformation process in which they combine to produce desired outputs. Although such a flow implies that initial plans have been developed concerning the allocation of inputs and the processes by which they are transformed into outputs, there is no control element at this point.

To introduce control, the rest of the schematic is required. Sensors must be developed which will measure attributes of the transformation process and attributes of the outputs. These measures in turn are transmitted to the comparer which compares them to the standards resident in the memory unit. If the results being achieved in the physical system conform to the standard, then the effector does not change the status of the inputs or transformation process. If they do not conform to the standards, then the effector makes the necessary corrections in terms of alternative resource allocation patterns and/or modifications in the transformation process itself.

If a physical system were to be developed along these lines the following definitions could be used:

Transformation Process: That activity which is productive and is to be controlled. It transforms inputs into outputs.

Sensor: A device which measures certain characteristics of the process and product being produced.

Comparer: A device which serves to compare the information forwarded by the sensor to the standard, specification, or desired characteristics of the process and product being produced.

Memory Unit: A device for holding information concerning the standard, specification, or desired characteristics of the process and product being produced.

Effector: A device which actuates some component of the equipment involved in the process or the allocation pattern of the inputs.

Many examples of mechanical, electrical, and hydraulic systems could be cited which illustrate these devices. A governor in a combustion engine, a thermostat in a heating system, and even the float valve in a toilet demonstrate the same control characteristics. In every case certain inputs undergo transformation processes yielding outputs. Some device must be established to measure the status of the transformation process and the outputs. These measures must be compared to standards and changes made through an effector when the standards are not being achieved.

The basic input-process-output scheme and associated feedback-loop concept used in the design of self-controlled mechanical, electrical, or hydraulic systems can be applied to management of the firm. Within the firm the four flow networks represent one approach to the development of systemic patterns in the physical system.

The real elements — materials, money, manpower, and energy — can be viewed as inputs to the system. They flow into the transformation process to create the outputs — products and services destined for the consumer. Machines and facilities also are characterized by flow patterns but their rates of flow are far slower than those associated with materials, money, manpower, and energy. Only infrequently does the firm acquire new machines from suppliers and dispose of old ones. Even more infrequently does the firm acquire or dispose of facilities.

During the transformation process, the materials, money, and energy are allocated in a manner which results in the production of products and services. These products and services can be viewed as the outputs of the system. They represent in another form, time, and place, the transformed inputs. Each product carries with it part of the material flow system in its physical makeup. Its modified form and position represent part of the machines, manpower, money, and energy expended on its conversion.

As examples of how such a transformation process can be measured, consider the production of a typical product. The input materials can be specified as to types, quantities, qualities, and so forth. After transformation, the resulting products can be measured in like terms. The machine requirements can be specified in terms of machine time required per unit giving consideration to the types of machines involved and their set-ups. Manpower can be measured in a similar way specifying skill levels and amount of time required per unit. In like manner, the energy flow and money flow required for successful transformation of the inputs can be determined.

Not all of the input flow of these elements necessarily emerges as physical outputs. It is easy to visualize the flow of materials from the input to output stage. But what about manpower, money, machines, and energy? By viewing these systemically it is possible also to visualize their flow patterns. The traditional view of them requires some modification, however.

Consider the input, manpower. Traditionally this is viewed as so many people of particular skill levels employed by the firm. Such a view results in a relatively static assessment of manpower in terms of a given number in the workforce. It would be just as realistic and far more useful systemically to view manpower in terms of man hours of various skill types available for the transformation process over time.

The flow process in this case can be viewed as a daily input flow of man hours which are expended in the transformation process.

Those that are used on a given day represent committed or allocated manpower resources. Those which are not used represent available manpower resources. Each day, this input stream of man hours is replenished with a new day's supply which flows through the system during the day.

The input resource, energy, operates in the same fashion. In the form of electrical energy it is metered as it enters the firm and its magnitude and rate of flow are measured. It is allocated as required to the transformation process by the use of switches and circuits which provide for its flow to the point of use. Gas, as energy, follows a similar flow pattern which is physically established in the transformation process by valves and pipelines.

The input resources, machines and facilities, also can be viewed in terms of flow if one considers that they have a useful "life." Over time they supply a flow of machine hours to the transformation process. After sufficient time this flow decreases for older machines and they reach the end of their productive "life." They then are removed from the firm to the environment and new machines with a given flow capacity and new "life" are brought in from the environment.

In all of these cases the inputs may be allocated as resources and committed to the transformation process. Such allocated resources at any given time can be said to be committed. Some of them, however, will remain uncommitted if the system is imbalanced or not operating at its full capacity. These resources would then be available, as are committed resources after their commitments over time are fulfilled. A schematic representation of this flow process appears in Figure 6-6.

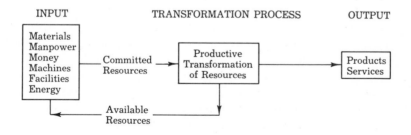

Figure 6-6.   Resource Flow Patterns

The key element involved in the design of a system composed of the four flow networks described is the determination of balance. In order to achieve this balance both the inherent capacities (volume) and rates of flow must be considered.

The machine capacity, facility capacity, energy capacity, manpower capacity, money capacity, and materials capacity define the outer limits of the capacity of the system. Given M_1 capacity of machines, M_2 capacity of facilities, M_3 capacity of energy, M_4 capacity of manpower, M_5 capacity of money, and M_6 capacity of materials, all expressed in units per time period T, it follows that the limit of the system is the minimum M_i.

In order to bring the system into balance it is necessary to modify the capacities M_i so that they match. Failure to do so will result in systemic imbalances characterized by such symptoms of problems as material shortages, machine bottlenecks, manpower limitations and so forth. This process in design, the balancing of capacity, is just one of two important aspects.

The other related design problem is one of balancing rates of flow among the elements in the flow networks. This presumes that the rates of flow of different specific elements (productivity of particular men, machines, etc.) may differ in the system but can be balanced by altering the sequence of flow patterns over time.

For example, if a particular machine produces 100 units per hour and the man who works with it can produce 50 units per hour, then the allocation pattern would call for the assignment of two men to the machine. Such assignment, however, requires study of the man-machine interface to design a machine capability of being able to receive 100 units per hour from two operators. In many instances this is not possible and thus the system tends to be limited in its effectiveness by the design properties of particular elements in it.

Assuming that both system volume and flow rates among the input elements can be balanced (admittedly a tenuous assumption given current concepts and conditions), then the next problem becomes one of resource allocation which will fully utilize the system over time. In addition, such resource allocation requires a delicate balancing of inputs among themselves and against capacities as output requirements change. In both the mechanical, electrical, and hydraulic systems which are self-regulatory, and in the firm as a system, this problem requires planning and control. A schematic representation of such a planning and control system appears in Figure 6-7.

This schematic is markedly similar to the one used in the design and operation of self-controlling mechanical, electrical, and hydraulic

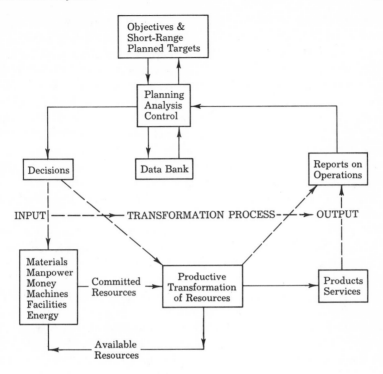

Figure 6-7.   Managerial Planning and Control System

systems. The significant differences are that there are many more variables to be considered which are ill-defined and that the planning and control system is an information system using symbols, not a real system using physical objects.

In today's larger business organizations, such a symbolic information system is required for integration and coordination of diverse activities. It is no longer possible for one man to plan and control his operations at first hand.

In place of the sensors found in engineered systems are management reports. These, in effect, report the status and attributes of the outputs of the system and also track the transformation process itself. Tracking outputs alone is insufficient since a feedback loop based on this will only provide corrective action when it is too late to affect the particular outputs from which readings were taken. By also tracking the transformation process, corrective action can be initiated in time to modify the process so that outputs will reflect the desired characteristics. The effectiveness of such a process, of course, implies that the feedback loop is executed very rapidly.

Another distinction between the feedback loop envisioned in the management planning and control system and that used in self-corrective mechanisms is that the feedback in the former system may occur periodically or continuously. In some cases, a total reading of the system status is taken periodically for analysis. In other cases, numerous readings are taken continuously but are only reported when the system status exceeds stated standards. This amounts to the utilization of management by exception.

In this age of computers and information retrieval it is possible to sense and report more information than a typical management team could read, let alone study, in one day. Thus, the massive data collection should be handled on an exception reporting basis with only periodic dumps of the data file taking place for intensive analysis of the complete status of the system.

The foregoing implies that a data bank is involved in the management information system. In the schematic this is shown as an accessible repository available to those involved in performing the planning, analysis, and control function. This data bank should contain the complete file for the status of the system both in terms of its transformation process activity and outputs produced at any given time. In addition, it should have the capacity to generate the exception reports mentioned above.

The development of the exception reports suggests the availability of standards for comparison. The repository for these standards is analogous to the memory unit in self-correcting systems. In it would be the stated objectives of the system as well as short-range targets defined by prior plans.

The comparer unit of the self-corrective system is replaced in the management information system by a group of people. These people are involved in performing the functions of planning, analysis, and control. It is their function to review exception reports and plan corrective action as the exceptions arise. This fast response capability is essential to the smooth operation of the system and avoidance of oscillation and system disequilibrium caused by delays.

In addition, they would prepare resource allocation plans for future time periods, both short- and long-range. This would require analysis of the state of the system with respect to its flow networks and their capacities. The short-range planning would result in directives which in turn would result in resource commitment to the transformation process. This is analogous to the effector in a self-corrective system.

The long-range planning would be designed to formulate strategies and consider risk factors as well as potential long-range problems which may create system disequilibrium. Although it would not result in directives on a daily basis, certain directives may be issued with respect to future problems associated with system imbalance such as modification of the size and nature of the manpower network or the number and type of machines in the machine flow network.

The heart of the management planning and control system described is the group of people associated with planning, analysis, and control. The effectiveness which can be expected from their efforts is a function of the knowledge of the dynamics of the information system of the firm as well as knowledge of the interfaces of the firm with objects in the environmental set.

Many of the skills required in today's manager are needed in such a group. These include business judgment, ability to assess and take risks, to weigh subjective factors which defy quantification, and to think in a creative and innovative manner. In addition to this, certain skills would have to be augmented. Ability to analyze the interaction of variables and constants in mathematical and statistical terms and the ability to use the computer effectively to assist in management decision making are two cases in point.

In addition to this, entirely new skills would be required. These skills would involve the ability to analyze the operation of the firm in terms of flow networks, to develop plans with respect to the interaction of these networks, and to utilize system simulation as a laboratory test of planning alternatives.

Organizational Consequences of a Systems Approach

Traditional management concepts such as line-staff, span of control, unity of command, and especially functionalism do not lend themselves readily to an organization in which the systems approach is utilized. Management, in the systems sense, involves decision making at critical decision points. It relies on the timely flow of pertinent feedback information to those points which have been sensed at the output and transformation process points in the physical system. The nature of an integrated management information system is such that it crosses functional boundaries and, when fully developed, replaces functionalism *per se*.

The information feedback pattern is not governed by hierarchical organization structures in such a manner that a report of an occurrence at a point in the physical system requires a number of approvals before it is passed "upward" through a chain of command or formal organization. Rather, the nature of the information feedback pattern is determined by the type of information to be transmitted and the type of decision required.

For example, a number of process control information feedback loops may only run from a part on a machine, through a sensor, to be checked against a standard automatically, and based on that comparison, an effector may or may not be activated to change the process. This is common in automated machines. A similar example concerns routine "management checks" where status reports are sensed from the transformation process or outputs, then automatically compared in a computer to particular management plans and standards and, if they are within limits, a programmed management "decision" will be forthcoming from the computer. If they exceed the limits, then an exception report is generated which will trigger the need for analysis and decision by particular managers whose expertise bears on that class of problems. Since a systems approach involves skills which cut across functional lines, such a group may require people with backgrounds in multi-functional areas, i.e., flow network specialists.

At periodic points in time the total status of the system will be reviewed by managers to ascertain potential systems design problems. The results of these reviews may take the form of additions or deletions of manpower, money, materials, machines, and facilities as the situation demands. Such decisions would require rigorous analysis and system simulation to test alternate resource allocation plans.

It is difficult to chart such an organization since it changes frequently to meet the needs of these information flows and decision points. Instead of one man being assigned to a given department, he may be assigned to a series of decision points as his skills apply. Instead of moving "up" in the organization through the chain of command, he may start by working with sensors and effectors, be promoted to the handling of exception report problems, and finally concern himself primarily with system design and long-range, balance problems. When properly implemented, such a promotional pattern must create generalists rather than specialists.

Note that the systems approach with its computer programs for handling routine decisions eliminates many responsibilities which are now held by middle managers. However, by removing most of the routine decisions, it provides more time for the managers involved in the system to study and analyze situations, to test alternate decisions prior to implementation, and to focus on long-range potential problems before they become today's brush fires.

The Major Subsystems of the Firm

From a systems point of view there is no consensus as to the nature of the major subsystems of the firm. As in the case of the relationships of the firm with the objects in the environmental set, too much is unknown with respect to *what* exists and *why* for theorists and practitioners to agree to a standard classification of subsystems.

There is theoretical agreement on the point that among the sciences, where systemic approaches have been taken in the past, each system seems to be made up of subsystems which are integrated in the larger systems. This belief, of course, can lead to detailed study of subsystems of subsystems of subsystems. Such an approach is, in part, an exercise in semantics, for what we view as a system may be interpreted by someone else as a subsystem of a larger system.

The usefulness of the subsystem concept lies in its simplicity. By breaking down a complex system into a myriad of subsystems it is possible to isolate a few inputs and outputs from the many in the complex system and study restricted transformation processes. Such simplification is often necessary simply to grasp mentally the input-transformation-output process involved in a particular case.

With respect to the firm as a system, it is clear that at the environmental level no model today exists which is operationally effective. As pointed out earlier, we simply do not know enough about the environmental relationships and why they operate over time as they do. Even at the firm level, the development of a level of systemic understanding which will yield optimal results has not emerged. Although important work is being done in the area of simulation at the firm level, we must restrict ourselves for the present to subsystems of the firm to discover operationally effective, integrated systems.

The subsystems of the firm could be designated in many ways. One might develop subsystems around the flow networks previously discussed. Or one might restrict the subsystem to areas in the firm where sufficient data are available to develop and operate integrated systems.

There is no doubt that the area of the firm which has the most hard data with which to work and which has used scientific approaches to the establishment of parameters and standards is that associated with operations in manufacturing. It is quite clear, in most cases, *what* exists with respect to a given manufacturing process, and *why* the process operates over time as it does. This reaches its highest form in those processes which have been automated. In such cases, the inputs can be easily identified, the transformation processes are sequential and carefully engineered, and the outputs are known. Relationships among capacities of the components in the process and flow rates are carefully balanced. Indeed, many of them have servomechanisms which cause them to be self-controlling.

As for other areas of the firm, such is not the case. To a greater or lesser extent the systemic relationships involved in such functional areas as marketing, finance, and personnel are less well defined. In accounting, effective systems work has been done but this is simplified by the fact that accounting is a symbolic information system which operates according to parameters which have been designed for purposes of simplification. That is, many accounting principles are adopted to provide uniformity and simplicity in the operation of the accounting system without meeting the test of a one-to-one correspondence with real-world phenomena, which the accounting system presumably tracks. Overhead accounting is a good example of this type of simplification.

The real challenge of systems analysis is not in the area of developing elaborate models and simulators which boggle the imagination in their sophistication. It is in effectively solving real-world problems so that organizations using it meet their objectives.

To illustrate how systems analysis could be applied to the development of an integrated planning and control system, we have chosen to define one of the subsystems of the firm as the inventory and materials management planning and control system. Through a fairly thorough analysis of the nature of this subsystem, its inputs, transformation processes, and outputs, its information system, and relevant decision points, it is hoped that some idea of the potential for effective application of systems analysis can be established.

Reference will be made to other subsystems within the operations sector of the firm in the discussion of the inventory and materials management planning and control subsystem. These five subsystems might be described as follows.

1. Production planning and control subsystem

The problems addressed in this subsystem include those associated with demand forecasting, overcoming fluctuations, balancing production, routing, scheduling, dispatching, expediting, control of operations, coordination of complex projects, and handling crash programs. Determination of such items as optimal production lot sizes, optimal product mixes, delivery dates, levels of production, lead times, and job priorities are also within the province of this subsystem.

2. Process and quality planning and control subsystem

The problems associated with this subsystem include setting process and quality standards, maintaining standards of uniformity and interchangeability, and meeting legal requirements. Some others are finding defective parts, finding problems in the process and correcting them, grading products, reducing scrap, and determining when, where and how much to inspect.

3. Cost planning and control subsystem

The problems in this subsystem involve cost minimization, cost estimating, determination of cost performance, differentiating between accounting costs and costs developed for decision making, opportunity costs, and problems associated with standard costs versus actual costs. Also associated with this subsystem are problems involving direct labor costs, indirect labor costs, materials costs, and overhead; cost and quantity relationships, effects of set-up costs and overtime costs, and the determination of production budgets.

4. Manpower planning and control subsystem

These problems are examined in this subsystem: determination of manpower requirements, job design, improving employee efficiency, setting performance standards, providing proper compensation, handling fluctuations in the workforce, overtime problems, holiday and vacation problems, and problems associated with manpower loading.

5. Facilities planning and control subsystem

The following problems are associated with this subsystem: evaluating needs for new machines and replacement machines, machine selection, effects on cash flow, depreciation, and taxes; determining plant capacity, and line balancing. Other areas include problems of automation, problems of layout, maintenance problems, buy-versus-lease decisions, and materials handling problems.

This tentative classification of subsystems in the operations area, with special reference to manufacturing which yields products, represents but one approach to the structuring of subsystems within the firm. In each subsystem the emphasis is on the integration of planning and control to find solutions to problems and maintain some equilibrium in the operation of the system. As such, this classification represents a point of view since the problem areas listed could be, and in fact are, classified in many other ways in various firms.

The difference in point of view can be significant, however. From a systems point of view it is clear that the decisions emanating from management must follow planning activity and that once they are executed there must be feedback for control purposes which, in turn, flows directly to the planning function. In short, planning and control from a systems point of view operate together to solve problems and maintain system equilibrium. Thus in each subsystem the adjectives planning and control are added to "system."

To indicate the significance of such a point of view, consider for the moment how management traditionally concerns itself with these problem areas. Generally management has concerned itself with production planning and control and established departments to handle both activities. But inventory management activities usually are concerned with people in "inventory *control*." The emphasis is clearly on keeping inventories from getting out of control, or fighting "brush fires" when they get out of control, rather than on effective inventory planning which might eliminate the "brush fire" hazard in the first place.

With respect to the activities in the cost area, groups are usually formed to handle "cost *control*." The concept of cost planning is just as significant yet that activity goes under the label of budgeting which, because of management emphasis on control in this area, still carries a negative, punitive stigma rather than emphasizing its positive opportunities.

Consider next the traditional designation for those in an organization concerned with manpower. If the organization has a department in this area, it is usually known as "manpower *planning*." While this is useful we must also address ourselves to the questions of manpower control — a concept which, under traditional jargon, almost offends. Yet from a systems point of view it is clear that the two go together.

The facilities area presents a similar case of short-sightedness. In a firm, this group may fly the banner "facilities *planning.*" Yet once facilities are planned it is necessary to provide for control information and feedback.

In the area of quality of the product, people consider themselves in the field of "quality *control.*" Yet before one can exercise quality control, standards, specifications, and control limits must be set through quality planning. This change of emphasis swings attention from just "meeting specs" to the consideration of what quality levels should be established in the first place.

There is no doubt that experienced managers in all of these areas do consider both the planning and control aspects of their jobs. Yet a change in point of view can provide more balance to the management activity. While managers of typical plants are comfortable with such concepts as inventory control, quality control, cost control, manpower planning, and facilities planning, they are somewhat uncomfortable with the counterpart concepts of inventory planning, quality planning, cost planning, manpower control, and facilities control. The validity of the planning activity matches that of the control activity in each case yet it is a bit disarming to consider how viewpoints can limit perspectives to the end that one-sided management emerges in these areas.

Summary

Four general areas are explored in this chapter. The first of these is the concept of flow networks. Four such flow networks are described: 1) the material flow network; 2) the money flow network; 3) the manpower flow network; and 4) the facilities, machines, and energy flow network. In addition, consideration is given to the role of the information system and the order flow network in the management of operations. The second general area concerns the analysis of the key planning and control variables associated with the management of the flow networks. The third area covers the organizational consequences of a systems approach to management. The fourth area explores a classification system of the major subsystems of the firm.

It is proposed that a meaningful understanding of the firm as a system must be based on empirical referents rather than on arbitrary classification systems developed along the lines of functionalism. Therefore, the four fundamental flow networks proposed in the systemic model are based on physical objects which are commonly found in business operations.

One class of such objects covers materials. It is suggested that the materials associated with the operations of the firm can be tracked from the vendor through manufacturing operations and distribution channels to the consumer. This flow pattern can be visualized as a network in which any change at any point in the network impacts operations at other points in the network. This being the case, effective planning and control requires an analysis of that continuum of activities which spans the material flow from vendors to consumers. Artificial segmentation of this flow pattern into organizational functions such as procurement, production, and marketing may lead to suboptimization for the firm.

The money flow network represents a continuous flow from several objects in the environmental set into the firm. These objects include customers, stockholders, and financial institutions. The money flow also moves from the firm to several objects in the environmental set including suppliers, stockholders, banks, labor, government organizations, and the community.

Manpower planning and control can be facilitated by conceptualizing a manpower flow network. In this case, the unit which is in flow is represented by particular hours of specified skills available for allocation as a resource over given time periods such as a day or week. During each of these time periods, manpower flows into the firm and is consumed in the transformation process. In an aggregate sense, one can also consider manpower flow with respect to the environment. In this case, the inputs to the firm come from the labor pool, competitive firms, and the community. Employees also leave the firm to return to the labor pool, government, the community, and competitive firms.

The machines, facilities, and energy flow network exhibits flow characteristics when viewed over time frames of different lengths. It is possible to visualize the flow of input energy to the firm and its routing through circuits to points of use and consumption in the transformation process. However, visualization of machine and facility flow networks requires expansion of the time frame. Particular machines flow into the firm from the environment and out of the firm to the environment. This amounts to acquisition of new machines and disposition of old or obsolete machines. In a shorter time frame, machine hours can be considered like man hours. That is, these machine hours can be allocated as resources on a day-by-day basis to the transformation process. Facility flow patterns are even longer than those of machines. New facilities are added and old facilities removed over time.

The function of the information system is to establish the appropriate sensors and units of measure for data acquisition required for the management of the four flow networks. In this capacity it serves as the primary interface between management and the empirical phenomena taking place in operations. The order flow network, although it exhibits flow characteristics through the firm, should be considered simply as a classification system rather than an empirical system.

The key planning and control variables associated with flow networks are discussed in this chapter with reference to a mechanical analog of a flow system. The discussion of this mechanical analog indicates that there are three fundamental variables: the capacity of the system, the volume in the system at any point in time, and the rate of flow through the system. Maintenance of equilibrium within the system requires feedback at certain specific check points. In addition to the consideration of volume, capacity, and rate of flow, it is necessary in complex organizations to consider convergence and divergence patterns as well as balance of rates of flow along these patterns.

To develop effective planning and control of flow processes, it is useful to adapt the feedback loop which is utilized so effectively in hydraulic, pneumatic, and electronic systems. Such feedback loops include sensors, comparers, memory units, and effectors. The translation of the engineering concept of the feedback loop to managerial purposes requires that 1) the sensor be considered as a series of reports on operations; 2) the comparer be replaced with a planning, analysis, and control group; 3) the memory unit be considered as the repository of objectives and short-range planned targets; and 4) the effector be considered as decisions emanating from the planning, analysis, and control group. In addition, some form of data bank must be established which is accessible to the planning, analysis, and control group. The operation of such a modified feedback loop provides the opportunity to utilize the management by exception principle. The planning, analysis, and control function is exercised in those cases where subjective considerations must be added to the objective analysis and logic embedded within the routine conditions.

The organizational consequences of a system approach radically modify the existing patterns of management in terms of such concepts as line-staff, span of control, unity of command, and functionalism. Many of the middle managers now involved in data reduction and data transmission functions are no longer necessary since these functions can be accomplished with computer hardware and software. The key decision makers are those involved in the planning, analysis, and control group. Although many minor decisions are automated, a number of exception decisions are handled in real time instead of being subject to the characteristic delays associated with hierarchical organization structures.

Although the four flow networks described in this chapter represent the fundamental concepts underlying the systemic construct, five other subsystems are denoted which are referred to in Part II of the book. These subsystems represent transitional definitions which closely reflect current views of the activities within the operations management sector. They include the following: 1) the production planning and control subsystem; 2) the process and quality planning and control subsystem; 3) the cost planning and control subsystem; 4) the manpower planning and control subsystem; and 5) the facilities planning and control subsystem.

In addition to these five subsystems there is the inventory planning and control subsystem which is associated with the material flow network. Part II of the book explores this area in detail and provides a frame of reference for exploration of systems analysis and system design in operations management.

Conceptual Problems

1. As a tentative classification of the flow networks involved in the firm, the following have been cited: 1) materials; 2) money; 3) manpower; and 4) machines, facilities and energy. Consider alternative classification schemes and additions to or deletions from these four flow networks.

2. The feedback-loop concept has been applied to the development of a managerial planning, analysis, and control function. In a theoretical sense, such a feedback loop could be considered closed. In reality, however, the objectives and short-range targets are not fixed but are constantly impacted from the environment. To develop a feedback-loop application in this case would require sensors of these impacts. An open-loop system would result and the system could be considered to be adaptive to its environment. How should these impacts be sensed and their forces measured?

3. What should be the characteristics of an effective manager in a firm using the systems concept? Why would they differ from the spectrum of skills required by managers in organizations using traditional management concepts?

4. An integrated systemic approach in a firm would require an integrated management information system. What are the hardware and software implications inherent in the design and operation of such an information system?

5. In this chapter the key variables capacity, volume, and rate of flow are discussed. Apply them in a dynamic analysis of the following:
    a. the materials flow network
    b. the money flow network
    c. the manpower flow network
    d. the machine, facility, and energy flow network.

6. Patterns of convergence-divergence are critical in systems design and systems analysis. Using a firm with which you are familiar, trace in general terms the convergence-divergence patterns associated with the following:
    a. the materials flow network
    b. the money flow network
    c. the manpower flow network.

Part II

Systems Analysis
and the Materials Flow Network

Inventory Planning and Control Subsystem

As is pointed out in Chapter 1, the design of this book is based on starting our investigation of the systems approach with broad, general systems concepts and then proceeding to deal with narrower systemic ranges. Figure 1-1 depicts this sequential approach.

In Part I, the general systems concept is examined in very broad terms and its application in various fields of science is explored. Then the systems approach is taken in an inquiry into the nature of the relationships to the firm of the objects in the environmental set. The impacts of government, customers, competitive firms, the community, labor unions, stockholders, banks, and suppliers are examined and evaluated. These objects impact the firm, are impacted by decisions made in the firm, and impact each other.

A continuation of the narrowing of the scope in the systemic hierarchy in Figure 1-1 leads to an examination of the firm as a system. Such a systemic view leads to the proposition that fundamentally the firm is composed of a series of flow networks. These include the materials flow network, the manpower flow network, the money flow network, and the machine, facility, and energy flow network. Such a conceptualization yields many new perspectives on the practice of management and executive decision making.

In Part II of the book we shall narrow the systemic focus even more to a detailed examination of the materials flow network and its component subsystems. These subsystems include inventory planning and control, inventory stock status, logistics, and purchasing.

In Part I, we are concerned primarily with broad conceptual matters. To introduce the systems approach at the level of the environment of the firm and at the level of the firm in its entirety will require much further research and experimentation before effective operative decision systems can be developed and implemented. In an effort to address the problem of pragmatic application of the systems concept, in Part II we shall apply it at a level where sufficient data exist and relationships are understood well enough to develop an operative system.

In essence, Part II is an application of systems analysis to the development of an operative materials flow network. Within the limited scope of this single network it is possible to integrate physical systems, management information systems, and decision systems to achieve organizational objectives.

As a brief overview of Part II let us consider the four subsystems in the materials flow network. As indicated in Figure 7-1, the materials flow network can be viewed as a linkage of four decision and information subsystems. These include the inventory planning and control subsystem, the inventory stock status subsystem, the logistics subsystem, and the purchasing subsystem. All four of these subsystems are linked to the physical system which is schematically represented as a sequence starting with vendors who supply inventory items, a receiving function, a raw materials inventory, an in-process inventory (which is, in fact, several intermediate buffer inventories), a finished goods inventory, a shipping function, and a final delivery to the customers of the firm. These points in the network of the physical system are, of course, linked by necessary materials handling activities.

The *inventory planning and control subsystem* is the analytical center of the materials flow network. It is responsible for determining forecasts of inventory requirements, determining the appropriate quantities of items to stock and to order, determining when orders should be placed, and maintaining appropriate levels of inventory turnover, among other functions.

The *inventory stock status subsystem* essentially performs an information gathering and disseminating function. Tracking of stock levels of items in raw materials, in-process, and finished goods inventories is a critical function of this subsystem. In addition, it is designed to track inputs on order and in transit from vendors and outputs from shipping to customers. Data gathered on stock status

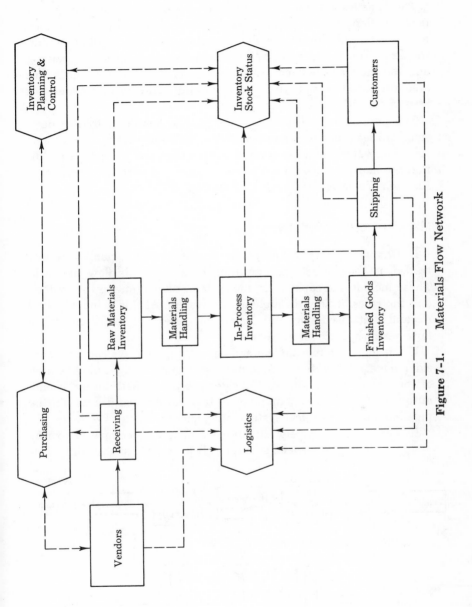

Figure 7-1.　Materials Flow Network

in this subsystem are, of course, essential for the decision processes taking place in the other three subsystems.

The *logistics subsystem* is designed to coordinate the transportation functions from vendors to the firm and from the firm to customers. It also coordinates the materials handing function within the firm. The interface between vendors and the firm at receiving and the interface between the firm and customers at shipping are also operated under the decision and information processes associated with the logistics system.

The *purchasing subsystem* is designed to handle analytical decisions in areas such as make-versus-buy, price determination, vendor selection, value analysis, systems contracting, and so forth. In addition the many information processes associated with the purchasing function are performed in this subsystem.

Material Flow Network Dynamics

In Chapter 6, certain key planning and control variables are discussed as they affect the operation of a materials flow network. These include the capacity in the system, the volume in the system, and the rates of flow throughout the system. At this point, we shall expand that discussion with particular reference to the materials flow network depicted in Figure 7-1.

Using the same mechanical analog in Chapter 6, we can now consider material flow network dynamics through a slight modification of the design of the analog. Consider a pipeline with an input end, output end, and three standpipes to take up fluctuations within the pipeline. Such a physical system is depicted in Figure 7-2.

In this physical system, standpipe A might be compared to raw materials inventory levels, standpipe B to in-process inventory lev-

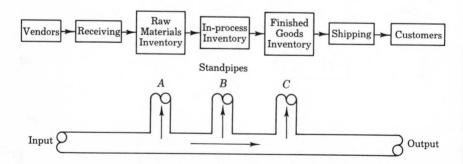

Figure 7-2.   Mechanical Analog of Materials Flow Network

els, and standpipe C to finished goods inventory levels. If fluid is introduced into the input end of the pipeline under pressure, it will tend to flow toward the output end of the system and may also back up into the standpipes when the lower pipeline is full. This is analogous to the situation in an industrial setting where inventories serve fundamentally as buffers for variations in rates of flow through the production process.

The open-ended standpipes are characteristic of inventories which, without control mechanisms, are subject to overflow conditions. The operating characteristics of this mechanical analog are simple and analogous to a real inventory system. At start-up, the pipeline fills from input to output end and then the overflow, if any, will fill the standpipes. The tendency of the system to overflow or become depleted is a function of two factors, the *volume* introduced into the system and the *rate of flow* of material through the system.

In order to operate the mechanical system effectively it is necessary to forecast expected requirements, which in turn determine the size of the pipelines to be used; to determine the quantities introduced into the system; and to control rate of flow from one position to another along the pipeline. This is analogous to the three fundamental models in the inventory planning and control subsystem: forecasting models, *EOQ* models, and reorder point models.

The schematic in Figure 7-2 has no provision for these factors and thus represents a poorly designed system from the point of view of planning and controlling operations. An improvement would be to introduce valves at the input and output ends of the pipeline. Consider the operation of the system with these additions, assuming that the initial state of the system is a filled pipeline and all three standpipes are half-full. Should the output valve suddenly open (analogous to a sudden surge in demand or usage rates) then the pipeline would drain the standpipes (various levels of inventory from finished goods back through the system to raw materials) and finally the input-output pipeline. To overcome this problem of depletion (stockout), it would be necessary to build in a sensor which would note the opening of the output side valve and actuate an effector which would very quickly open the input side valve an equal amount. With real-time response capabilities it would be possible for the system to react sufficiently fast so that depletion of the pipeline would not occur. In many business situations, unfortunately, the information feedback loop from the output side (customers) to the input side (vendors) is so slow that by the time corrective action is taken on the input side (placement of orders for raw materials) the status of the output side valve has changed.

To follow this through, suppose that the output side valve now closes suddenly. Immediately the pipeline fills and the standpipes overflow. The only corrective action which would prevent this would be a real-time feedback response to the input valve to close it as well. Since such changes in rate of flow and quantity are characteristic of industry and since feedback response time is rarely real time, is it any wonder that most inventories fluctuate rather substantially as the oscillation patterns build over time?

The analogy can be expanded somewhat at this point to differentiate among types of inventories, i.e., raw materials, in-process, and finished goods inventories and standpipes *A*, *B* and *C*, respectively. With only input and output valves and a sensor at the output end of the pipeline, any surge through the system or drop in quantity or rate of flow will affect equally the inventory level in all three standpipes. Since this is not realistic as an industrial analogy it is necessary to insert more valves and sensors as indicated in Figure 7-3.

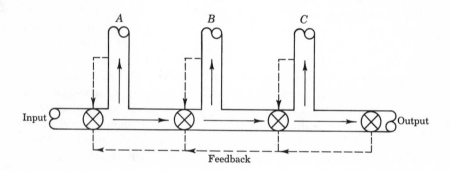

Figure 7-3.   Mechanical Analog of Materials Flow Network

In Figure 7-3 two additional valves have been inserted between standpipes *A* and *B* (raw materials and in-process inventories) and between standpipes *B* and *C* (in-process and finished goods inventories). With appropriate feedback loops from one valve to another and sensors connected to each standpipe the system now has the flexibility to take up imbalances in three areas internally, standpipes *A*, *B*, and *C*. This is characteristic of a feature which is valuable where subjective factors require modification of inventory levels independently, rather than level and balanced throughput. For example, it might be advisable to temporarily overload raw materials in view of an impending strike faced by a supplier of critical materials. Or it might be prudent to temporarily overload finished goods inventories if customer service requirements are such that throughput time cannot satisfy them and servicing from finished goods is the only way to meet competition.

Even though such imbalances and limited independence of operation are built into the system, the valves in the system should be linked so that the impact of change in a position downstream is fed to control mechanisms upstream in time to prevent system instability. A final problem to be faced is that which occurs when rate of flow or quantities exceed the valve and pipeline capabilities. That is, overflow is inevitable. Such a situation is depicted in Figure 7-4.

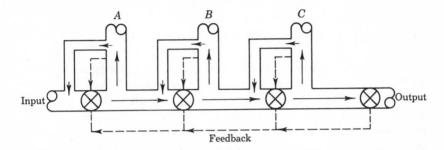

Figure 7-4. Mechanical Analog of Materials Flow Network

Figure 7-4 illustrates the case of a system which has overflow bypasses from one standpipe to a position prior to its preceding valve in the main pipeline. This is analogous to the control over turnover and dead stock. Standard operating inventories should operate as in Figure 7-3. However, over time some items in inventory will accumulate to create excessive levels of non-working inventory. When this occurs the items should be examined for disposition. In Figure 7-4 the finished goods are diverted back for rework in the process. The in-process inventories are diverted back to salvage as raw materials. Other disposition decisions, of course, might divert excess inventories out of the system entirely which is tantamount to venting the system or scrapping the materials in an industrial setting.

This mechanical analog illustrates the fundamental relationships associated with the planning and control of an inventory system. It should be noted that the key elements are control over *volume* in absolute terms and control over *rate of flow* between successive points. The analog illustrates that inventory management fundamentally is a function of planning and controlling quantities and the movement of items from station to station over time. It is all too easy to fall into the trap of dollarizing the relationships among variables and thus losing touch with the fact that the managerial planning and control system must replicate what is, in fact, a physical system of flowing materials.

Thus, in system design in this area, it is important to measure events in units and time which are real value systems and reserve the use of accounting in dollars to those situations where the variables involved cannot be expressed or homogenized in any other way. Accounting conventions in the inventory area too often reduce the facts of the physical inventory system to fiction. And, unfortunately, this fiction becomes the interface for the managers whose decisions affect the operation of the physical system.

The Inventory Planning and Control Subsystem

The inventory planning and control subsystem is designed primarily for analytical decision processes. The outputs of these processes significantly affect the other subsystems — stock status, logistics, and purchasing. Although the systemic relationships among the four subsystems are symbiotic, the decisions made in the inventory planning and control subsystem have the most significance in terms of the potential effectiveness of the materials flow network.

The analytical decision processes involved in the inventory planning and control subsystem include the following:

1. Development of forecasts of expected inventory requirements for each inventory item.
2. Determination of the appropriate quantities to be ordered and carried in inventory.
3. Determination of when orders for inventory items should be placed to minimize stockouts.
4. Maintenance of appropriate levels of inventory turnover with control over slow-moving items and dead stock.

Each of the analytical decision processes above will be discussed in further detail in addition to the necessary system interfaces to identify the appropriate inputs and outputs to the respective processes. Prior to these discussions, however, attention should be given to the various types of inventory planning and control systems which might be utilized in any given situation.

Types of Inventory Planning and Control Systems

In the design of any inventory planning and control system there are three important input variables which must be considered —

time, quantities, and costs. Several types of systems have been designed which relate these variables; a few types are described below.

The simplest system is the so-called two-bin system. In this system, costs and time values are secondary considerations. The basic variable used for control is quantity. The fundamental idea in this quantity-based system is that a regular stock is maintained with an emergency stock available to signal the need for replenishment. These relationships are depicted in Figure 7-5.

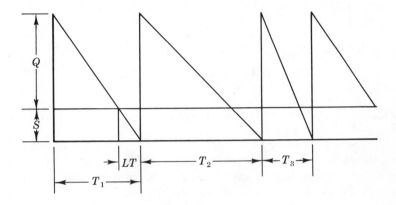

Figure 7-5. Quantity-Based Inventory System

Given a regular stock of Q units and an emergency or safety stock of S units, the system operates in the following manner. The initial state of the system is $Q+S$ units in inventory. Over time, the units in the regular stock, Q, will be used in production processes. When Q is depleted and stock must be taken from the safety stock S, then a replenishment order is placed through the purchasing subsystem.

It is assumed that the quantity S will cover the necessary lead time LT for procuring materials, including such activities as making requisitions, selection of vendors, preparation of purchase orders, follow-up, receiving and inspection of the replenishment stock, and placement of the stock received in inventory. Note that dollar limits are not directly applied in this system and that the period of time consumed to use Q units is a variable rather than a constant. Quantity is the only sensed signal in the feedback loop for replenishment.

Although this system is quite simple to install and requires the least amount of computation on the part of inventory personnel to use, it is subject to a rather critical limitation. The safety stock, S, is set at a level which assumes a constant lead time and a constant usage rate. Changes in either of these could lead to overstocking or stockouts. This condition is depicted in Figure 7-6.

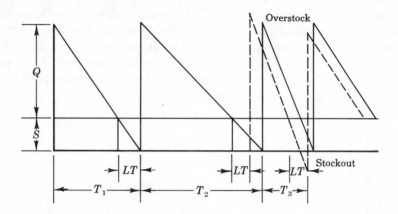

Figure 7-6. Effect of Variable Usage Rate and Constant Lead Time

Assuming that the quantity $Q + S$, divided by the time period T_1, represents an expected usage rate, and LT is assumed to be the necessary replenishment lead time and remains constant for time periods T_1, T_2, and T_3, then in the time period T_2 where the usage rate decreases a slight overstocking will occur. In time period T_3 where the usage rate increases, the lead time is insufficient to prevent a stockout. A replenishment order of Q will bring total inventory levels to less than those which existed during time periods T_1 and T_2.

With this system, in conditions of oscillating usage rates or oscillating lead times, occasional stockouts and overstocking may occur in an oscillating fashion representing moderate system disequilibrium. However, in successive periods of increasing usage rates or increasing lead times or both, moderate disequilibrium of the system is quite likely to become amplified wherein stockouts will occur regularly and with greater severity as the rates of usage or lead times continue to increase. The opposite type of system amplification could occur with the opposite states of the two critical system variables leading to overstocking although this is somewhat dampened by the replenishment decision process based on depletion of inventory levels to the safety stock S.

Since this inventory system lacks sufficient feedback to control overstocking or stockouts, other systems have been developed to check these conditions of system disequilibrium. One reasonably common alternative is a time-based system wherein, at periodic time intervals, the inventory level of the item to be controlled is checked. This check may be made through an information system relying on a perpetual inventory record or may be made through inspection of the physical inventory. In many cases the time-based inventory system has been developed as a by-product of accounting conventions which required the information for periodic weekly, monthly or quarterly reports rather than for optimal operation of the inventory system. Such a system is depicted in Figure 7-7.

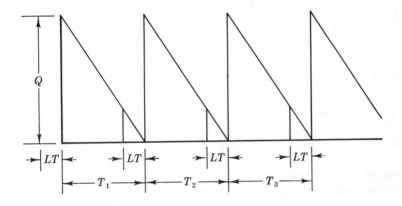

Figure 7-7. Time-Based Inventory System

In the time-based system, the time periods T_1, T_2, and T_3 are equal — a week or month, for example. In Figure 7-7, the system state at the outset indicates full stocking of inventory at the beginning of time period T_1. At the end of T_1, the inventory level is checked and a replenishment order is placed for Q units which arrive after the lead time LT has expired in period T_2. At the end of period T_2, another check is made and the inventory is again replenished by Q units after the lead time LT extending into period T_3 has expired and so on.

Although in this system the important variable of time is considered rather than neglected as in the prior system, it is still subject to overstocking or stockouts should the lead time or usage rates change. An example of this can be seen in Figure 7-8.

In Figure 7-7, the time-based inventory planning and control system is stable due to the constant lead times and the constant usage rates. In Figure 7-8, however, a condition of overstocking occurs over time and is amplified from Q to Q_1 to Q_2. The stimulus to the

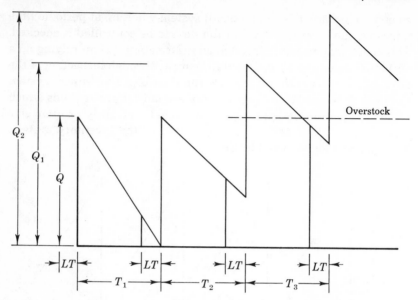

Figure 7-8. Overstock Effect of Decreased Usage Rate

system which caused this was a reduction in usage rate in a system characterized by constant time intervals, constant lead times, and constant replenishment quantities. This same condition could occur when lead times are decreased. Most firms, which have anticipated this situation or learned of the amplification tendency through experience, have modified the time-based system by adding a maximum-inventory-level restraint to check overstocking. Under these conditions the order placed at the end of period T_2 would be the

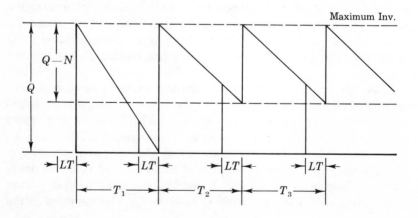

Figure 7-9. Damping Effect of Maximum Inventory Constraint

maximum minus the projected stock status at the end of the lead time or $Q - N$. Such a constraint on overstocking, used in conjunction with a time-based system, is depicted in Figure 7-9.

Now let us turn our attention to the conditions of accelerated usage rates or increased lead times in a time-based system. Figure 7-10

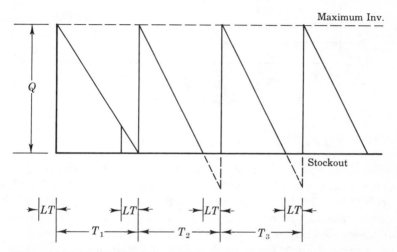

Figure 7-10. Stockout Resulting from No Minimum Constraint

reveals that increased usage rates alone may result in stockouts. If lead times also increased, the stockout condition would be amplified. Note that although the time-based system provides periodic review with emphasis on the system variable, time, it does not have a damping factor protecting against stockouts as does the "two-bin" system. Thus, an out-of-stock condition may exist and not be noticed until the next periodic inventory check. If it is noticed prior to the inventory check, it usually is brought to the attention of those responsible for management of the materials flow network in the form of a rush order. This situation is a characteristic of management by crisis which the systems concept attempts to minimize.

One approach to the solution of stockouts, with time-based systems, is the addition of a safety stock to absorb effects of accelerated usage rates or increased lead times. To cover the necessary lead time for replenishment, it is also useful to establish some minimum inventory level which serves to signal the need for replenishment activity. These sensors, when combined with a maximum inventory level constraint, provide the elements of what is commonly known as the min-max system. Figure 7-11 illustrates the characteristics of such a system.

The min-max system provides constraints against overstocking and protection against stockouts. The system, therefore, becomes

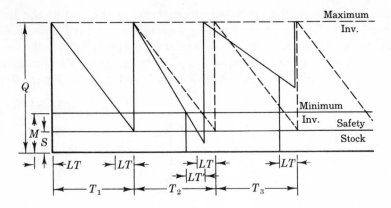

Figure 7-11. Stockout and Overstock Constraints

homeostatic and reasonably stable. The only real difficulties with such a system concern its effectiveness in terms of cost and service capabilities. The following questions require answers. What is the optimal quantity to purchase of any inventory item? When should orders for replenishment be placed to safeguard the firm from unreasonable stockouts? How much safety stock is necessary to provide adequate protection without incurring excessive costs for this protection?

Although these questions are often answered through the use of judgment, experience, and intuition, a great deal of study has been expended to provide objective decision processes to find reasonable, if not optimal, answers. Several subsystems, designed to provide these answers, will be examined in some detail in succeeding chapters.

Development of Forecasts for Inventory Planning

In order to plan and control inventories, it is necessary to develop some estimate of the future requirements which the various inventories are to serve. Some inventories, such as small buffer in-process inventories between two machines which produce at equal and stable rates, can be planned with little trouble for there is limited uncertainty about variations in demand. Other inventories including finished goods inventories of a fashion or fad nature may require much subjective judgment to determine future patterns of demand. Still other inventories, like those associated with many raw materials, involve some element of uncertainty, yet there remains sufficient stability with respect to particular patterns of demand, that statistical techniques can be effectively used to forecast demand. Other inventories may not require forecasts of any sophistication at all.

The examples cited above illustrate the point that the development of forecasts for inventory planning should not be built around *one* sophisticated and all-encompassing model in a computerized decision system. The types of forecasts and forecasting techniques which are appropriate depend on the types of inventories involved and how they operate within the firm as linking devices to other transformation components of the system as well as the environmental set.

In the cases cited above, the in-process inventory forecast might be built simply in terms of the space available to convey parts between the two machines and the speed at which they move. In this sense the quantity required over time would be constant and the dynamics of the situation would correspond directly to the ejection rate of the preceding machine and the feed rate of the following machine.

Inventories of a fad or fashion nature, particularly those in a finished stage which cannot be economically modified if the market demand pattern changes, require forecasts of an entirely different type. In this case a forecast might be derived from solicitation of opinions from customers, retailers, wholesalers, salesmen, competitors, and various other sources ranging from trade journals and other publications to the pronouncements and performances of purveyors of what is "in" in the fashion world. Needless to say such forecasts are general and subjective and are described in terms of executive judgment, skill, foresight, insight, hindsight, and any other kind of sight which penetrates the murky fog of market uncertainty. A little luck proves very helpful when forecasts of this type must be made.

The raw materials inventory situation cited as an example might be a prime candidate for the application of a variety of statistical techniques to the forecasting process. These might include moving averages, least squares regression, exponential smoothing, lead-lag analysis, various cyclical approaches, or Monte Carlo simulation, among other quantitative approaches. Since these approaches are appropriate for computer applications and systemic forecasting, they will be explored in greater depth.

The final example is intended to focus attention on the observation that many items in any list of inventory items do not justify expensive control techniques or fancy forecasting models. The lowly paper clip and rubber band are classic examples of items in this type of inventory.

A useful approach to the separation of inventory items for forecasting, planning, and control purposes is the *A-B-C* concept shown in Figure 7-12. The *A-B-C* idea reflects the importance of considering the value of inventory items in relation to the total array of items in particular inventories.

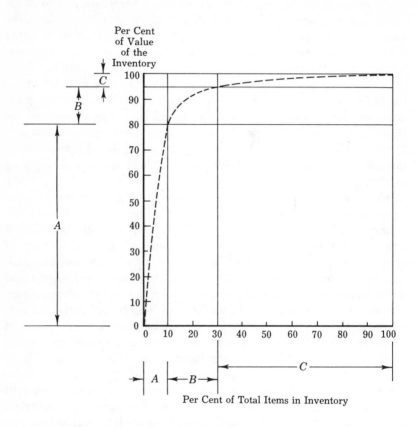

Per Cent of Total Items in Inventory

Figure 7-12.   Value Categorization of Inventory Items

Just as forecasting techniques should vary according to the type of inventory item in question, consideration should also be given to the relationships of value to numbers of items in inventory. In Figure 7-12, a hypothetical yet rather typical example of many firms, a small number of total items in inventory represent a very large proportion of the total value of the inventory. The items thus categorized as *A* items represent only 10 per cent of the items but account for 80 per cent of the value of the entire inventory.

Such items are usually critical to the efficient and economical operation of the company. They represent the bulk of the usage in most cases and the usage rate, typically, is rather rapid. Given these conditions, such inventory items justify careful planning and control. Considerable study can be given to finding effective and economical answers to the questions of when to order and how much to order. In addition, fairly elaborate forecasting techniques are justifiable since the savings involved in effective inventory management of these items are substantial.

The items categorized as B items in Figure 7-12 represent 20 per cent of the items in inventory but only 15 per cent of the value of the total inventory. These items are considered important for consideration in terms of inventory planning and control as well as forecasting; however, the forecasting techniques and planning and control efforts over these items should not be as expensive or elaborate as those applied to the items categorized as A items.

The final category of inventory items, noted as C items in Figure 7-12, represents the bulk of the items in the inventory (70 per cent) but only a small proportion of the value of the total inventory (5 per cent). Item C includes such things as the paper clips and rubber bands mentioned earlier. Also represented are those slow-moving items for which demand has slackened; those boxes of items in the warehouse on which the purchasing agent got such a great quantity discount five years ago but can't seem to use up; the scrap items people refuse to classify as scrap because "you never know when we can use the stuff," and the like. These sorts of inventory items, plus many of low value and moderate usage which fit the C category, justify little effort in the way of inventory planning and control and less of formalized forecasts of inventory demand.

Although this cost categorization scheme separates inventory items into three categories, this is simply a tradition. In particular applications, it may be appropriate to develop further segmentation to differentiate among inventory items. Such differentiation is a process of focusing attention on the fact that inventory planning and control and forecasting of demand cost money. This money should be allocated in such a manner that important inventory items are carefully managed yet not all items are subjected to costly analysis.

Such analysis is also useful in providing categorization of items for the routine operation of the inventory stock status subsystem. Different routines of managerial action are used with each category and such categorization serves as an important branch of data entering and leaving the inventory stock status subsystem.

Summary

In this chapter the materials flow network is examined. The management information system and decision system associated with it involve four related subsystems. The first, the inventory planning and control subsystem, is designed to serve an analysis function yielding answers to such questions as what future requirements will be, how many items should be manufactured or ordered from vendors, and when production and purchase orders should be placed. The second, the inventory stock status subsystem, is designed to track volume within the system at various physical points and at particular intervals of time. The third, the logistics subsystem, is designed to track rates of flow at particular points in the system at specific time intervals. The fourth, the purchasing subsystem, is designed to provide the necessary liaison between the firm and vendors in its environmental set.

Material flow network dynamics are also discussed which reveal some of the relationships of capacity, volume, and rate of flow as they affect planning and control activities within the materials flow network. The fundamental requirements for effective planning and control are brought out by tracing requirements for the operation of a mechanical analog of a materials flow network.

Two basic types of inventory planning and control systems are examined. The first, a quantity-based system, is workable but subject to disequilibrium when impacted by either changing lead times or changing usage rates or both. The second, a time-based inventory system, considers the time variable and is a commonly encountered system. It is still subject to overstock or stockout, however, should usage rates change substantially.

The introduction of maximum and minimum inventory constraints as well as a safety stock to buffer system shock induced by accelerated usage rates and increased lead time largely overcome these problems.

Finally, consideration is given to the question of forecasting as it is employed in the inventory planning and control subsystem. Various types of inventories require different types of forecasting techniques. In addition the use of A-B-C analysis suggests that value and usage of items make a significant difference in the selection of the particular forecasting techniques which should be employed for specific inventory items or classes of items.

In Chapter 8 we shall consider a variety of forecasting techniques which can be introduced in the decision system utilized within the inventory planning and control subsystem. These techniques include least squares regression, moving averages, and exponential smoothing.

Conceptual Problems

1. Considering the schematic depicting the inventory and materials management subsystem described in this chapter, attempt to translate the conceptual scheme into practice by using it in conjunction with a company with which you are familiar. Identify the convergence patterns from vendors to receiving and divergence patterns from shipping to customers. Examine the convergence-divergence patterns which exist from raw materials inventory to finished goods inventory. Where would the logical sensor points exist for logistic tracking and stock status monitoring?

2. Determine the types of inventory items which could be controlled effectively with a quantity-based inventory system. What are the limiting assumptions which affect your choice of appropriate inventory items?

3. Determine the types of inventory items which could be controlled effectively with a time-based inventory system. What are the limiting assumptions which affect your choice of appropriate inventory items?

4. What factors should be considered in establishing stockout and overstock constraints in an inventory system? How can the magnitudes of these values be analyzed using dollars, quantities, and other measures? Where do probabilistic impacts enter the analysis? Should the maximum and minimum levels be re-evaluated periodically? If so, how often should this occur?

5. *A-B-C* analysis serves to distinguish among types of inventory items for purposes of planning, analysis, and control. Since this separation of inventory items into three categories is traditional, the question arises, Should further differentiation be used to separate items into four or more categories? What factors, other than value and usage, should be considered in such a classification scheme?

6. A mechanical analog of a material flow network is presented in this chapter. It emphasizes the importance of considering capacity, volume, and rate of flow in system design. The same concept can be applied to manpower, money, and machine flow networks. Conceptualize how such flows could be integrated since they necessarily impact each other over time.

Forecasting Materials Flow
Network Requirements

In this chapter we are concerned with the determination of inventory requirements. A forecast of demand, translated into requirements for specific time intervals, is the initial process in the decision system utilized in the inventory planning and control subsystem. All other decisions stem from this important information input. The mathematical model used to determine how much to order, the economic order quantity to purchase or manufacture, requires this information. In turn, the mathematical model used to determine when orders should be placed, the reorder point model, requires quantity requirements as an input.

Since different inventory items require and justify different types of forecasting techniques, a number of them are discussed in this chapter. Initially, we shall consider the differences between stable demand, trends, and cyclical demand patterns. Then consideration will be given to a number of statistically oriented forecasting techniques including the computation of the least squares regression function, averages, moving averages, weighted moving averages, and exponential smoothing. In addition, certain techniques designed specifically to track cyclical demand patterns will be investigated. These include analysis of the ratio of cumulative demand to actual demand as well as the comparison of last year's demand and average demand to develop a cycle factor for adjusting forecasts for cyclical patterns. Finally we shall consider the problem of forecast error. Obviously, since forecasts represent extremely critical inputs to

8

the information and decision systems, errors in them can create a great deal of system disequilibrium. To deal with the problem we shall examine the use of MAD or mean absolute deviation.

Forecasting Techniques

The variety of forecasting techniques which have been developed over the past several years is substantial. They range from the unknown mental processes of the experienced executive whose "intuition" and "feel" for the market is uncanny, through the collection and consideration of various people's opinions, to mathematical and statistical techniques that occasionally become so complicated that the researcher who develops them cannot communicate them to anyone else and achieve a reasonable level of human communication and understanding.

In this chapter, a few of the basic techniques will be examined briefly to convey how the development of forecasts for inventory purposes could be systematically approached and interfaced with the other components of the inventory planning and control subsystem. Three conditions of demand will provide useful foci for discussion: relatively constant demand patterns, positive or negative trend patterns, and cyclical patterns, particularly those associated with seasonality.

Figure 8-1 depicts a relatively constant demand pattern such as might be expected of in-process inventories between balanced machine operations functioning at reasonably stable levels. In such a case, a forecast can be made using the arithmetic mean which is a measure of central tendency wherein the observations are summed and divided by the number of observations. In such a case the value

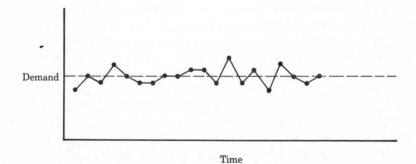

Time

Figure 8-1.   Stable Inventory Demand

of the dependent variable (inventory level) is an assumed constant over the range of the independent variable (time).

Although there is some fluctuation about the mean in Figure 8-1, as is typical of most stable inventory demand conditions, these fluctuations are randomly distributed. There is no discernible trend in demand either upward or downward and no cyclical pattern.

Figure 8-2 represents a condition of positive trend in inventory demand. Again there are fluctuations above and below the dashed line which indicates the linear trend. These fluctuations are random, yet, over the twenty periods plotted, there is a steady increase in demand, which is not random.

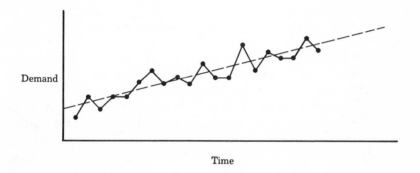

Time

Figure 8-2.   Trend in Inventory Demand

In order to project expected demand into the future in cases of trend, it is possible to utilize moving averages, exponential smoothing, least squares regression, and other less elaborate techniques. All of these approaches are designed to extrapolate historical trend into the future. Such projections are usually in error to some degree due to random fluctuations. The development of a sound approach to forecasting trends in inventory demand, therefore, requires both a projection technique and a measure of forecast error.

The third common demand pattern for inventory items is seasonality or some other reasonably short-frequency cyclical pattern of demand. This condition is depicted in Figure 8-3 where demand is plotted for quarters of a year at the points January, April, July, and October, over a span of six years. It is apparent from these plots that the July plot for each year reflects peak demand every year. The base line or dashed line in Figure 8-3 serves to highlight these periodic fluctuations and also indicates that there is little or no trend in the data.

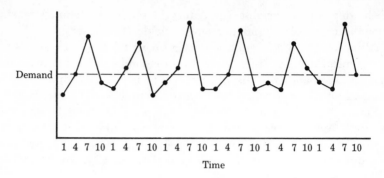

1 4 7 10 1 4 7 10 1 4 7 10 1 4 7 10 1 4 7 10 1 4 7 10

Time

Figure 8-3.   Seasonality in Inventory Demand

Although seasonality implies that the cyclical phenomenon occurs with a periodicity of twelve months, other cycles may also be present in inventory demand patterns. These may be short cycles of a few weeks duration or longer cycles of several years duration. Regardless of the frequency of the cycle, it is important that the forecasting technique chosen for cyclical phenomena measure both the frequency and amplitude of the fluctuation as well as the magnitude of forecast error. In some cases, it also is necessary to examine each of several cycles which are operating simultaneously in the market and build the forecast on a composite of the interrelationships among the various components of demand at a given future point in time. Such a set of cycles is depicted in Figure 8-4.

Note that the one-year cycles in Figure 8-4 are quite regular and represent seasonal variation, peaking in the summer and reaching annual troughs in the winter. Viewed by itself, the seasonal cycle for the year ahead would not be difficult to forecast since its amplitude per year remains fairly stable.

The two-and-one-half-year cycles depicted in Figure 8-4 also show regular patterns of variation. When Figure 8-4 is examined closely, it can be seen that the pattern for each period is almost identical and, if the same conditions were to hold for the coming year, it would not be difficult to establish the forecast based on the pattern shown in the two-and-one-half-year cycles.

The composite cycle is the combination of both the one-year and two-and-one-half-year cycles at fifty points in time over five years. It reveals a pattern which appears to be very peculiar over the time

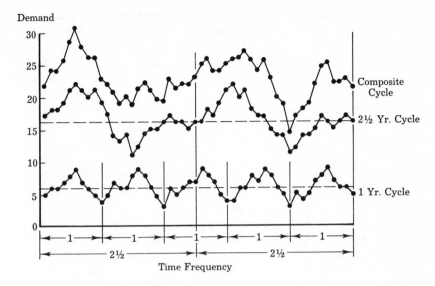

Figure 8-4.   Composite Cycle Components

span of five years. Although this composite bears little resemblance to the two clear cycles below it in Figure 8-4, it too would represent a cyclical pattern over a longer period of time.

This shorter duration was chosen for illustrative purposes to indicate that a rather odd pattern as indicated by the composite cycle may be made up of very regular sub-cycles of differing frequencies and/or amplitudes. In this case the frequencies work together and then against one another over the five-year period. For example, in year one, the positive slopes of both component cycles reinforce each other to create extreme peaking of the composite cycle. Later, in year two, the one-year cycle peaks as the two-and-one-half-year cycle reaches its trough; thus they cancel out each other and yield a composite cycle which appears to be fairly constant over the second year.

The third year contains the typical one-year cycle but its peak is only slightly amplified by the slow concurrent rise in the two-and-one-half-year cycle. At the beginning of the fourth year, the peak of the two-and-one-half-year cycle and the trough of the annual cycle again offset each other, indicating stable overall demand in the composite cycle. During most of the third and fourth years, therefore, the total demand represented by the composite cycle appears to be stable although its components are fluctuating rather widely. If one were to predict, in the middle of the fourth year, that the gentle increased trend, indicated from the second year through the middle of the fourth year, would continue, there would be a rude awakening in store.

During the latter part of the fourth year, the troughs of both the two-and-one-half-year cycle and the annual cycle correspond, thus yielding a precipitous drop in demand in the composite cycle. This is just the opposite of what might be expected if one viewed only the composite cycle through the middle of the fourth year. It is, however, the result of two perfectly regular sub-cycles operating simultaneously on different frequencies.

In the fifth year both sub-cycles reach peaks simultaneously which accounts for the rapid recovery from the precipitous drop of the fourth year. Although the explanation of the behavior of the composite cycle in Figure 8-4 is clear when the interaction of the two sub-cycles are known, in most cases one starts with raw data which look much more like the composite cycle than the sub-cycles. Then the problem becomes one of analysis of the composite cycle to find relevant seasonal cycles or other cycles of shorter or longer periodicity.

To determine the component cycles which make up a composite configuration, it is necessary to review several years of past data and determine cause-effect relationships among those determinants of cyclical change. Some of the causes of cyclical demand may be generated internally by the firm as a result of its purchasing, production, or marketing policies. An interesting analysis of these internal cyclical relationships is represented by the excellent work of Forrester.[1]

Other determinants of cyclical phenomena are external to the firm. Some have to do with patterns associated with the industry of which the firm is but one component. Others have to do with aggregate economic cycles. Still others reflect cyclical patterns in certain demographic characteristics of the population.

To this point we have examined briefly three types of demand patterns which often apply to the management of inventories. These types include reasonably constant demand patterns, trends, and cycles. In each of these cases, a number of forecasting techniques

---

[1] J. W. Forrester, *Industrial Dynamics* (New York: John Wiley & Sons, Inc., 1961).

have been developed which are useful for analytical purposes. These are discussed briefly in the remainder of this section of the chapter.

Least Squares Regression

The least squares regression technique is designed to fit a linear function through a scatter diagram. In forecasting, such a scatter diagram represents the plots of demand over several time periods. As illustrative examples consider Table 8-1 and Figure 8-5.

TABLE 8-1

Least Squares Regression Computations

Year	x	Y	xY	x^2
1	−5	10	−50	25
2	−4	11	−44	16
3	−3	12	−36	9
4	−2	13	−26	4
5	−1	12	−12	1
6	0	13	0	0
7	1	13	13	1
8	2	14	28	4
9	3	14	42	9
10	4	15	60	16
11	5	16	80	25
$N=11$	0	143	55	110

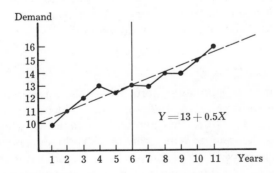

Figure 8-5. Least Squares Regression Function

One approach to the determination of the linear function which represents the assumed trend is to determine the equation of a line

$(Y = a + bx)$ where the Y axis and Y intercept (a) occur at the midpoint time period. In the case of the data in Table 8-1, the Y axis occurs at the year 6.

To determine the value of the Y intercept and the slope (b), the years are listed in order; the deviations from the midpoint year are noted as positive and negative integers (where an odd number of years are involved), as indicated in the table under the heading "x"; the demand values are listed under Y; the product of the deviations (x) times the demand values are listed; and a final set of values, the deviations squared (x^2) are noted. In cases where an even number of time periods appear in the table, the positive values are noted as $+0.5$, $+1.5$ and so forth and the negative values are -0.5, -1.5, et cetera.

Once the table values are determined, the columns are summed and the number of years is noted (N). This will yield values for N, Σx, ΣY, ΣxY, and Σx^2. The equation of the line can then be determined using the following two equations:

$$a = \frac{\Sigma Y}{N} = \frac{143}{11} = 13$$
$$b = \frac{\Sigma xY}{\Sigma x^2} = \frac{55}{110} = 0.5$$
$$Y = 13 + 0.5x$$

Since x represents deviations from the midpoint year, if a forecast for year 12 is desired, the result would be $Y = 13 + (0.5 \times 6)$ or 16. Similarly, if a longer range forecast were to be made for year 16 then the projection would yield $Y = 13 + (.05 \times 10) = 18$.

An alternative approach to least squares regression involves the solution of the following two equations:

$$\Sigma Y = Na + \Sigma Xb$$
$$\Sigma XY = \Sigma Xa + \Sigma X^2 b$$

This alternative is especially useful where the forecast is a function of an independent variable other than time which may not be serial or continuous. To utilize this approach, a table of values is established listing the independent variable (X), the dependent variable subject to the forecast (Y), the product of each set of independent and dependent variables, and the independent variables squared. In addition, the number of sets of dependent and independent variables are noted (N). Application of the same figures used earlier results in Table 8-2.

TABLE 8-2

Least Squares Regression Computations

X	Y	XY	X^2
1	10	10	1
2	11	22	4
3	12	36	9
4	13	52	16
5	12	60	25
6	13	78	36
7	13	91	49
8	14	112	64
9	14	126	81
10	15	150	100
11	16	176	121
66	143	913	506

$N = 11$

The equation of the line which represents the least squares regression function is determined as follows:

$$\text{Eq. 1} \quad \Sigma Y = Na + \Sigma Xb$$
$$143 = 11a + 66b$$

$$\text{Eq. 2} \quad \Sigma XY = \Sigma Xa + \Sigma X^2 b$$
$$913 = 66a + 506b$$

Multiplying Eq. 1 by six eliminates the term a as follows:

$$913 = 66a + 506b$$
$$-858 = -66a - 396b$$
$$\overline{55 = \qquad 110b}$$
$$0.5 = b$$

To solve for the term a, the value 0.5 for b is substituted in Eq. 1.

$$143 = 11a + (66 \times 0.5)$$
$$110 = 11a$$
$$10 = a$$

The equation of the line determined by this method is $Y = 10 + 0.5X$. The slope remains the same as before, the only difference being that the Y intercept is now based on Y at X equals zero rather than Y at the midpoint of the years as is the case with the first method.

If a forecast for year 12 is desired, the result would be $Y = 10 + (0.5 \times 12) = 16$. Similarly, a longer range forecast for year 16 would yield $Y = 10 + (0.5 \times 16) = 18$. The results, of course, correspond to those arrived at with the method discussed previously since both sets of answers are based on the same input data and the use of least squares regression as the approach to the approximation of trend.

Moving Averages

Averages can be used in several ways to develop forecasts. The simplest approach is to compute the arithmetic mean of a set of historical values. The equation used to determine the mean is $\Sigma X/N$. In the case of the data in Tables 8-1 and 8-2 the demand values noted as Y are treated as the X values for computation of the mean. Therefore the arithmetic mean would be 143/11 or 13.

This value represents a forecast based on the average of all values of X and completely overlooks any indication of cycle or trend. Since a definite trend exists in the data over time, the forecast of 13 is clearly too low for future periods.

One approach to the tracking of trend over time, while retaining the use of the arithmetic mean, is to employ a moving average. Using this technique, the arithmetic mean is based on recent data and changes over time as new data are added and old data are deleted from consideration. This tracking feature can be illustrated with the data in Tables 8-1 and 8-2.

Consider time periods 1 through 5. If a five-month moving average is employed, the projection would be $(10 + 11 + 12 + 13 + 12)/5$ or 11.6. The next computation would occur after period 6 and would include the data for that period while deleting the data for period 1. The result would be $(11 + 12 + 13 + 12 + 13)/5$ or 12.2. In periods 7, 8, 9, 10, and 11, the same procedure would be used basing the mean on a set of five figures which moves through time as new data become available. To demonstrate the tracking capability of the moving average, consider the five most recent figures. The result would be $(13 + 14 + 14 + 15 + 16)/5$ or 14.4. This result is closer to the indicated projection of the trend than is the figure of 13 derived from an arithmetic mean of all data.

Although the moving average provides a forecast which is more heavily weighted by current data than older data, it still is subject to a lag. The more time periods included in the moving average, under conditions of an increasing or decreasing trend, the greater that lag will tend to be.

One way to minimize the lag is to use fewer time periods. In the case of time periods 7 through 11, a five-month moving average gives a weight of 20 per cent to the latest figure as follows:

Period	Demand	Weight	Weighted Demand
7	13	.20	2.6
8	14	.20	2.8
9	14	.20	2.8
10	15	.20	3.0
11	16	.20	3.2

Forecast $= 14.4$

Using only the three latest time periods would yield the following forecast where the latest figure is weighted 33.33 per cent:

Period	Demand	Weight	Weighted Demand
9	14	.333	4.662
10	15	.333	4.995
11	16	.333	5.328

Forecast $= 14.985$

The forecast of 15 (14.985 is slightly in error due to rounding) is closer to the current data reflected by the trend than is the 14.4 figure resulting from a five-time-period moving average. This may seem to be an improvement; however, the chance of forecast error is greater when the number of time periods included in the moving average is decreased. An indication of the severity of this problem appears in Table 8-3.

The determination of the ratio in Table 8-3 is based on random demand periods where the standard deviation of the total error is equal to $\sqrt{1 + (1/N)}$ (where N is the number of time periods) times the standard deviation of demand.

Thus, a one-month moving average involves a forecast error which has a standard deviation 40 per cent larger than the standard deviation of demand. The error drops rapidly as the number of time periods in the moving average increases.

The choice of the number of time periods to include in the moving average is, therefore, one made under conditions of a dilemma. If too many time periods are chosen, the forecast will lag and not be very responsive. If too few are chosen, the potential forecast error increases. In this situation one might rely on longer periods where

TABLE 8-3

Expected Accuracy of Moving Averages*
(Ratio of the standard deviation of forecast errors
to the standard deviation of demand)

Number of Months in Moving Average	Ratio of Standard Deviations
1	1.414
2	1.225
3	1.153
4	1.118
5	1.095
6	1.086
7	1.068
8	1.058
9	1.054
10	1.049
11	1.044
12	1.039

*R. G. Brown, *Statistical Forecasting for Inventory Control* (New York: McGraw-Hill Book Company, 1959), p. 31.

conditions are reasonably stable and on short periods where demand patterns seem to be changing rapidly.

Another approach would be to select several different time periods and simulate the tracking characteristics and forecast errors when applied to actual demand patterns taken from past records. After such tests, a time period for the moving average, to be applied to particular inventory items, could then be selected for use.

One compromise, which partially overcomes this dilemma, is to utilize a weighted average where recent data are weighted more heavily than older data and where the increase in weights is not requisite on a decrease in time periods as is the case discussed earlier. For example, it was shown that a five-period moving average would result in 20 per cent weights being applied to each of the last five demand values. By establishing a non-uniform weighting scheme, it would be possible to increase the emphasis on more recent data while maintaining five time periods.

As an example, if the last five time periods of Tables 8-1 and 8-2 are used and the following weights are applied, the forecast strongly reflects the most recent data even without a reduction in the number of time periods.

Period	Demand	Weight	Weighted Demand
7	13	.05	0.65
8	14	.15	2.10
9	14	.20	2.80
10	15	.25	3.75
11	16	.35	5.60

Forecast $=14.90$

In this example the forecast increases from 14.4 to 14.9 when the 20 per cent equal weighting of the demand data in the five time periods is abandoned in favor of a weighting scheme which puts heavier emphasis on recent data without reducing the number of time periods.

Exponential Smoothing

The idea developed above with respect to heavy weighting of recent data is more fully developed in the concept of exponential smoothing. Exponential smoothing overcomes some of the problems associated with using the moving average as a forecasting technique. It was shown above that to maintain stability in a forecast, or to filter out the effects of random noise or fluctuations in demand data for past periods, it is necessary to use a fairly large number of time periods in the analysis. Coupled with the requirement of maintenance of long demand records for inventory items is the requirement that these figures be moved over time, adding recent data and deleting older data.

Exponential smoothing minimizes these problems by using only the forecast figure, a current actual demand, and the alpha factor for the inventory items for which demand is being forecast. This greatly simplifies record keeping and speeds the computation of the forecast, an advantage of some significance even when computers are used in those cases where hundreds of items require frequent analysis.

Although the weighting of a moving average can be changed by increasing or decreasing the number of time periods or by arbitrarily assigning particular weights, these methods of changing weighting patterns are cumbersome to handle computationally. Exponential smoothing provides a single weighting factor, alpha, which is easily manipulated and, in most cases, serves as well as other selected weighting patterns.

The basic concept of exponential smoothing is that one should adjust a forecast for the next time period bearing in mind the difference between the forecast for the last period and the actual demand of the latest period. For example, if actual demand for the last period exceeded the forecast for that period then the new forecast should reflect this and be somewhat higher than the old forecast.

The question is, How much higher? This is where α (the alpha factor) comes in. It is essentially a weight selected by management which indicates what percentage of the difference between the actual demand last period and the old forecast should be added to the old forecast to bring it up to date for the next period. In a sense it is a factor which updates the forecast in light of what has happened most recently to actual demand. Putting the concepts into equation form would yield the following:

New Forecast = Old Forecast + α (Actual Demand − Old Forecast)

For example, if the old forecast was 100 units, the actual demand was 120 units, and the alpha factor 0.10 or 10 per cent, then the new forecast would be 102 units as follows:

$$\text{New Forecast} = 100 + .10 \ (120 - 100)$$
$$= 102$$

An alpha factor may have a value between 0 and 1.0. Zero is excluded because that would amount to no change in the original forecast over time; an alpha of one is excluded because that would result in 100 per cent weighting of the difference and yield a forecast equivalent to current demand, thus losing any smoothing characteristics.

The impact of the difference between actual demand and the old forecast can be weighted very lightly, such as 0.01, 0.05, or 0.10 (1 per cent, 5 per cent, or 10 per cent), or very heavily such as 0.30, 0.40, or 0.50 (30 per cent, 40 per cent, or 50 per cent), or any other weighting desired. Due to the decay characteristics of the presumed influence of demand data for past periods which is inherent in exponential smoothing, a practical range of choice lies between 0.05 and 0.30.[2]

The use of a high alpha factor such as 0.30 results in adding significant weight to the difference between actual demand and the old forecast. Such a factor provides quicker response to changes in actual demand over time than is the case with a lower alpha factor. The presumed influence of older data also diminishes rapidly.

---

[2]C. McMillan and R. Gonzalez, *Systems Analysis* (Homewood, Ill.: Richard D. Irwin, Inc., 1965), p. 221.

The use of a lower alpha factor such as 0.10 lends only moderate weight to the difference, provides greater stability in the forecasts over time, and the presumed influence of older data stretches over a longer period.

The presumed influence of older data is a reflection on the shape of the distribution resulting from the application of a given alpha factor. The shape is that of an exponential curve. To demonstrate this the following notation is used: [3]

$F_n = $ Forecast for time period n
$D_n = $ Demand for time period n
$\alpha = $ Value of the alpha factor
$n = $ Time period; $F_0 = $ original forecast, $F_4 = 4$ periods later

Considering F_4 as the forecast to be made at the present time, weightings will be traced through periods 3, 2, and 1 back to the original forecast. The exponential smoothing equation is as follows:

$$F_n = F_{n-1} + \alpha (D_n - F_{n-1})$$

Verbally, this amounts to the present forecast being equal to the forecast of the prior time period, plus alpha times the difference between the actual demand this period and the forecast of the prior time period. To illustrate the derivation of the exponential function it is more convenient to rearrange the terms as follows:

$$F_n = \alpha D_n + (1 - \alpha) F_{n-1}$$

The forecast for each of the four prior periods, when viewed separately, would be as follows:

$$F_4 = \alpha D_4 + (1 - \alpha) F_3 \qquad \text{(Eq. 1)}$$
$$F_3 = \alpha D_3 + (1 - \alpha) F_2 \qquad \text{(Eq. 2)}$$
$$F_2 = \alpha D_2 + (1 - \alpha) F_1$$
$$F_1 = \alpha D_1 + (1 - \alpha) F_0$$

where $F_0 = $ original forecast estimate of demand.

In Eq. 1 it was noted that

$$F_4 = \alpha D_4 + (1 - \alpha) F_3$$

Substituting Eq. 2 for F_3 in Eq. 1 yields

$$F_4 = \alpha D_4 + (1 - \alpha) \ [\alpha D_3 + (1 - \alpha) F_2]$$
or $\qquad F_4 = \alpha D_4 + \alpha D_3 \ (1 - \alpha) + (1 - \alpha)^2 F_2$

Working back in time over all periods in like manner yields

$$F_4 = \alpha D_4 + \alpha D_3 (1 - \alpha) + \alpha D_2 (1 - \alpha)^2 + \alpha D_1 (1 - \alpha)^3 + (1 - \alpha)^4 F_0$$

---

[3] *Ibid.* See also pp. 218-219.

Factoring out alpha and expressing the model in terms of n yields the following general model:

$$F_n = \alpha[D_n + D_{n-1}(1-\alpha) + D_{n-2}(1-\alpha)^2 + \ldots D_1(1-\alpha)^{n-1}] + (1-\alpha)^n F_0$$

This model indicates that the weights given to demand decrease in an exponential manner as the data are older or farther removed from the present. A schematic illustration of this is shown in Figure 8-6 using three different alpha factors.

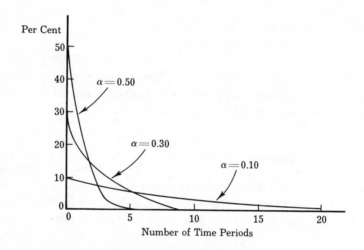

Figure 8-6. Exponential Curves with Three Alpha Factors

By comparison with moving averages, it becomes clear that the choice of the alpha factor affects the weighting of current data over older data. High alpha factors are comparable to a moving average over a very few time periods and low alpha factors reflect weightings similar to those obtained over longer periods. Table 8-4 reflects the alpha factor which compares with the number of months required in an equivalent moving average. These values are computed from the following equation: [4]

$$\alpha = \frac{2}{n+1}$$

Similarly Table 8-5 reflects the corresponding number of months in a moving average at selected values of the alpha factor. These data are based on the following equation:

$$n = \frac{2-\alpha}{\alpha}$$

---

[4]R. G. Brown, *op. cit.*, pp. 57-62 and Appendix B.

TABLE 8-4

Values of the Alpha Factor Corresponding to Selected Time Periods
in a Moving Average

Number of Months in Moving Average	Alpha Factor
3	0.500
4	0.400
5	0.333
6	0.286
7	0.250
8	0.222
9	0.200
12	0.154
18	0.105
24	0.080
36	0.054
48	0.041

TABLE 8-5

Number of Time Periods in a Moving Average Corresponding
to Selected Alpha Factors

Alpha Factor	Number of Months in Moving Average
0.01	199.0
0.05	39.0
0.10	19.0
0.15	12.0
0.20	9.0
0.25	7.0
0.30	5.7
0.35	4.7
0.40	4.0
0.45	3.4
0.50	3.0

Thus far we have examined the characteristics of exponential smoothing as a forecasting technique under conditions of fairly stable demand patterns. Other patterns result in different tracking characteristics, however. Three of these are discussed below: impulse, step, and trend.[5]

An *impulse* is a non-recurring increase or decrease in demand which occurs at one point in time over a range of time periods which otherwise exhibit reasonable stability. Such an impulse is a signifi-

---

[5]For further development of these concepts see *ibid.*, Chapter 2.

cant change in demand level rather than a small random fluctuation. In such a case, where an unexpected change impacts the inventory system, the response to the stimulus using exponential smoothing is a change in the forecast level (high or low magnitude dependent on the choice of alpha factor) followed by a gradual return to the equilibrium state.

A large alpha factor will result in a sizeable response to the impulse and a rapid return of the forecast to the equilibrium state whereas a small alpha factor will result in a smaller response and a slower return to the equilibrium state. This relationship is depicted in Figure 8-7. Note that the scale is discontinuous to enlarge the response pattern.

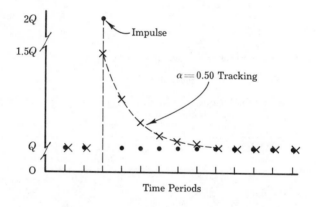

Figure 8-7.   Response Pattern to an Impulse

A *step increase* reflects an increase or decrease in demand which occurs at a point in time followed by a relatively stable demand pattern at the new level. Since the demand pattern does not return to the former equilibrium state but rather reflects a new average over time, exponential smoothing will respond to the change and then subsequent forecasts will increase over time approaching the new level.

As in the case above, the response pattern, in terms of magnitude of response and speed of correction of the forecast, is dependent on the choice of the alpha factor. This condition is depicted in Figure 8-8 where again the scale is broken to amplify the response pattern schematically.

Although exponential smoothing responds rather well to impulse and step changes in demand, as well as random fluctuations around

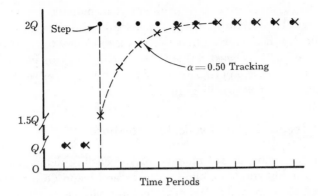

Figure 8-8. Response Pattern to a Step Increase

stable demand, it does not directly correct for demand changes due to *trend*. Just as a moving average will lag demand data under conditions of trend, so does the exponential average.

Since exponential smoothing detects change and adjusts the new forecast by adding a fraction of the difference between the current demand and the old forecast to the old forecast, under conditions of trend the tracking characteristics are such that it cannot keep pace with changes in demand along the trend (assuming the use of typical alpha factors). This condition is depicted in Figure 8-9 as the single-smoothing response.

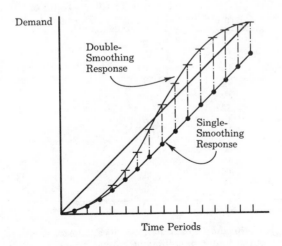

Figure 8-9. Response Pattern to Trend

In order to overcome this problem it is possible to use double smoothing.[6] This technique will track trend and compensate for it. One difficulty, however, is that there is a tendency for oscillation of double-smoothed forecast values. The oscillation can be minimized by selection of the appropriate alpha factor. The effect of double smoothing is also shown in Figure 8-9.

Forecast Techniques for Cyclical Fluctuations

Up to this point, useful models for forecasting demand for items with reasonably stable demand patterns and trend demand patterns have been examined. The cyclical demand pattern, particularly that associated with seasonality, requires a modified approach. One of the useful techniques, in this case, is to use indices of demand for each month in the year. Of course there may be cases of weekly demand cycles within each month. In such cases the technique could be applied using a week as a time period to develop useful cyclical forecasts.

As a starting point, the monthly sales patterns are examined over two or more past years. Based on these data, average sales figures for each month can be computed which will minimize the noise in the

TABLE 8-6

Ratios of Cumulative Demand to Annual Demand

Month	Demand	Cumulative Demand	Ratio of Cumulative Demand to Annual Demand
January	5	5	.02
February	10	15	.06
March	15	30	.12
April	20	50	.21
May	20	70	.29
June	30	100	.42
July	40	140	.68
August	50	190	.79
September	20	210	.87
October	15	225	.94
November	10	235	.98
December	5	240	1.00

[6]See *ibid.*, pp. 65–70, and C. McMillan and R. F. Gonzalez, *op. cit.*, pp. 226–227, for a discussion of the double smoothing technique.

demand history. On the basis of the average sales figures, projections can be made for future time periods. One approach to the making of such projections is to use the ratio of cumulative demand for a particular month to the annual demand and to apply this to the current demand pattern. The figures in Table 8-6 provide a frame of reference for discussion of the technique. The relationships in Table 8-6 are depicted in Figure 8-10.

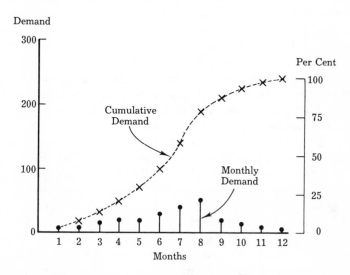

Figure 8-10.   Monthly and Cumulative Demand

In Table 8-6 the monthly demand figures show a marked seasonality, rising in the summer months and peaking in August. This seasonal pattern appears schematically in Figure 8-10. To develop forecasts for future months these monthly demand figures are totaled cumulatively by month and then a ratio of cumulative demand to annual demand is computed for each month. By multiplying current demand for any given month times the quotient of the ratio of the future month, divided by the ratio of the month corresponding to that for which present data are available, a forecast is generated for the future month in terms of cumulative units at that point in time.

In general, the idea behind this method is that seasonality will reflect a particular cumulative demand curve as shown in Figure 8-10. By knowing the current demand in a given month, it becomes possible to project the demand for future months and for the year based on knowledge of how this year's demand is running compared to the pattern of demand of other years. An example will clarify this concept.

If the average demand pattern is reflected in Table 8-6 and actual demand data have been gathered for the current year through the month of March (the first quarter), then it is possible to compute expected demands for the remaining portion of the year. For example, if a forecast were required for the next month, April, then the ratio for April would be divided by the ratio for March (0.21/0.12) yielding a factor of 1.75. This factor would then be multiplied by the current March cumulative demand, say 36 units, an increase of 6 units over the average of 30 at the end of the first quarter. The product of the cumulative demand to date and the factor computed above (36 units × 1.75) would yield a forecast of 63 cumulative units for the year at the end of April, or an April forecast of the difference, 27 units, between the cumulative forecast at the end of April and the cumulative actual at the end of March.

Utilizing this concept over a somewhat longer time span, one might forecast the demand at the end of June by dividing the ratio for June by the ratio for March (0.42/0.12) and multiply the resulting factor (3.5) times the March actual cumulative demand (36) to arrive at an end of June cumulative forecast of 126 units.

This approach is a mathematical analogy of the typical approach a businessman might use if he subjectively said to himself, "Knowing how demand goes up and down regularly over the year and knowing where peak demand hits and also having actual results before me at the end of the first quarter, I will predict that demand for future months and the year as a whole will follow a similar pattern; and I will adjust this pattern upward somewhat based on the actual increase in demand in the first quarter this year compared with the same period last year."

Another approach to forecasting seasonal demand patterns for particular months is based on applying the ratio of demand for a given month to average demand and multiplying it by the average monthly demand forecast in the coming year. Data in Table 8-7 illustrate this technique.

From the data in Table 8-7 one can see that average demand must be computed from the past year's demand (or averaged over several years to minimize noise, in which case the monthly demand data should also be averaged). By dividing each month's demand figure by the average, a factor can be developed each month to indicate the percentage of average demand each month reflects. This approach yields weights which reflect the inherent seasonality in the data

TABLE 8-7

Ratios of Monthly Demand to Average Demand

Month	Last Year's Demand	Average Demand	Demand/$\overline{X}$ Factor	This Year's $\overline{X}$ Demand	Monthly Forecast
January	5	20	0.25	28	7
February	10	20	0.50	28	14
March	15	20	0.75	28	21
April	20	20	1.00	28	28
May	20	20	1.00	28	28
June	30	20	1.50	28	42
July	40	20	2.00	28	56
August	50	20	2.50	28	70
September	20	20	1.00	28	28
October	15	20	0.75	28	21
November	10	20	0.50	28	14
December	5	20	0.25	28	7

Total Last Yr. 240   Total Expected This Yr. 336

which can be applied to the average associated with a new aggregate forecast to arrive at monthly forecasts.

In the case of Table 8-7 the past year's demand average was (240/12) or 20. It is presumed that an aggregate forecast for the next year yields an annual demand of 336. By dividing 336 by twelve months the presumed average monthly demand for next year appears to be 28 units. This presumption would yield a stable demand pattern with zero slope if left at this point. To reflect the seasonality inherent in the situation, the average demand for each month must be adjusted upward or downward by the factors indicated in the table to arrive at seasonally adjusted monthly averages. Such a set of seasonally adjusted averages appears in the last column of the table.

Exponential smoothing can be used in forecasting seasonal demand just as it is used in smoothing other data. As discussed before, this requires comparing actual demand characteristics with forecasts and applying a smoothing constant or alpha factor to achieve the level of smoothing desired.

Estimating Forecast Error

Developing a forecast for future demand is a problem which must be solved to insure effective inventory planning. On the basis of the forecast, future production plans can be laid and future inventory decisions can be made, particularly those dealing with the question of how much of a particular inventory item should be procured. Chapter 9 will show that the determination of annual requirements is a requisite input to economic order quantity decision rules and analytical techniques.

Determination of a forecast alone, however, is not sufficient for effective inventory and materials management. In conjunction with the forecast, it is necessary to determine actual errors and estimate expected forecast errors. The estimate of error is useful in determining safety stock levels which are part of the subsystem associated with reorder points. These reorder points are used in developing decision rules to answer the question, When should an order be placed?

Figure 8-11 has been developed to emphasize the importance of tracking actual forecast errors and estimating potential forecast errors. In this figure two cases are presented. Both Case I and Case II yield identical forecasts based on least squares regression; however, the reliance which can be placed in the forecast of Case II is con-

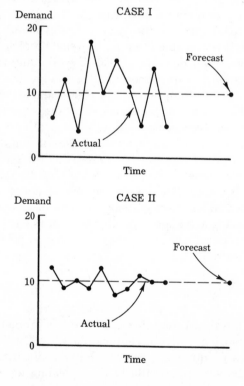

Figure 8-11.   Impact of Forecast Error

siderably greater than that which can be placed in the forecast of Case I. More importantly, the potential stockout problems of Case II are fewer than those associated with Case I.

Forecast errors, like deviations from the mean in a quality control sense, tend to cluster about the mean in a normal distribution in most cases. A useful statistic to measure the degree of clustering, or nearness of the data to the mean, is the standard deviation. The standard deviation is defined as the square root of the variance, and the variance, in turn, is defined as the arithmetic mean of the sum of the squared deviations of the observed values from the mean.

In a normal distribution, 68.27 per cent of the observed values will be within ± 1 standard deviation of the mean; 95.45 per cent of the observed values will be within ± 2 standard deviations; and 99.73 per cent of the observed values will be within ± 3 standard deviations of the mean. Thus we have available a statistic which allows for prediction of error distribution based on past data. This predictive ability lends itself to the analysis of levels of service which will be maintained, safety stock levels, and the determination of significant changes in the mean which are not due to chance.

By inspection, one can see that the standard deviation associated with Case I in Figure 8-11 is larger than that of Case II. On the basis of this, it is likely that larger safety stocks would be provided in Case I and that high levels of customer service might be difficult to maintain in Case I.

The procedure for developing the standard deviation requires tabulating all of the observed values, determining the arithmetic mean of these values, subtracting the mean from each of the values to determine deviations, squaring all the deviations, determining the mean of the squared deviations, and taking the square root of this mean. Needless to say, for a large number of inventory items with demand histories of moderate length, the computational problem becomes large scale. To simplify the computation of a statistic which effectively measures error, the mean absolute deviation (MAD) is often used.

MAD is simply the arithmetic mean of the forecast errors, disregarding sign. It is as useful as the standard deviation in measuring dispersion about the mean and requires much less calculation. Therefore it will be used as the measure of dispersion in the remainder of the discussion here. The $MAD = 0.8$ standard deviations (or 0.7978σ, more accurately); conversely, one standard deviation $= 1.25$ MAD, assuming a normal distribution. On the basis of this approximation of the standard deviation, forecast errors can be tracked and safety stocks set at desired levels of customer service.

In order to track forecast errors, it is possible to minimize even further the number of calculations required by using the exponential smoothing concept. In this case the following equation will serve to update the MAD forecast:

New MAD Forecast = α Actual Error + $(1-\alpha)$ Old MAD Forecast

The operation of the alpha factor on response sensitivity is similar to the characteristics discussed earlier. By tracking the history of forecast errors over time using the MAD forecast, it is possible to detect non-random variation about the mean. In most cases one would expect the range of errors to be rather small but when the sum of the errors begin to accumulate on the positive side of the distribution, or on the negative side, it is time to review the forecasting model. In such circumstances the actual mean may have shifted and the forecast is not effectively compensating for this shift. One remedial course of action may be to increase the alpha factor in the exponential smoothing equation for the forecast to quicken response. Another, perhaps more prudent course of action, should the sum of the errors shift suddenly and distinctly, is to review the actual data which have recently accumulated and try to discern subjectively what has happened in the market or in production operations which would so significantly effect dispersion from the forecast.

To build effective tracking into the model it is useful to carry along the sum of the errors. Once the magnitude of the errors crosses an upper or lower control limit, management is alerted to the fact. Where control limits are set tightly, this may create too many "alerts" due to random system oscillation. In such a circumstance, if control limits are not increased it is possible to institute a test for successive positive or negative sums of errors of particular magnitude which would indicate a shift of mean due to assignable causes rather than chance. Only when a particular number of positive or negative errors of this magnitude had accumulated would the information system alert management.

Summary

Forecasting models are perhaps the most significant models in the inventory planning and control subsystem. They represent the initial information input to several other decision models. Upon the forecast rest decisions relative to how much of an item should be purchased or manufactured and when these activities should take place. In addition, the forecast of demand has a significant impact on decision systems associated with the allocation of resources in flow networks other than the materials flow network.

The scheduling of machines and facilities is dependent on the forecast. The allocation of manpower is similarly affected. In addition, the money flow rates and volume requirements at particular points over time in the money flow network are impacted by the forecast and scheduling decisions derived from it.

In this chapter we examined several models which can be utilized in the decision system encompassed within the inventory planning and control subsystem. Least squares regression analysis, which yields a linear function, is suitable in cases of relatively stable or trend forecasts where the variance is nominal. The use of various averages is also considered, ranging from the simple arithmetic mean to moving averages. Since moving averages sometimes lag current activities, consideration is also given to the weighted moving average as a device to emphasize more current data.

Since a weighted moving average involves much computation in moving from one time interval to another, consideration is also given to exponential smoothing, a technique which is quite adaptive in tracking current demand yet retaining some past history. This technique has, among other advantages, relative simplicity in the computational phase and an easily adjustable weighting factor.

The problem of forecasting cyclical patterns is also examined. Two approaches are proposed: first, the analysis of the ratio of cumulative demand to actual demand; second, the comparison of last year's demand and average demand to develop a cycle factor for adjusting forecasts for cyclical influences. Finally the problem of estimation of forecast error is confronted. The technique proposed to solve this problem is the use of the mean absolute deviation (MAD).

In Chapter 9 we shall consider the mathematical models which can be used in determining how much should be ordered or manufactured. This decision is the one which follows forecasting in the decision system encompassed in the inventory planning and control subsystem.

Conceptual Problems

1. Several analytical planning and control techniques are discussed in this chapter. A critical question which arises when they are considered for use in industry is whether the value of the information generated by them exceeds the cost of analysis and operation of the system. This is particularly evident if operations involve building a

software package for computer processing. How would you develop a cost-versus-value study to determine the economic feasibility of using a particular technique? Examine all factors pertinent to such a decision and decide how their cost and value coefficients could be determined and presented to management.

2. Consult some books on forecasting and attempt to develop a useful statistical approach to forecasting demand which incorporates trend, seasonality, and cycles. Express the technique as a mathematical model and utilize it over several time periods. Then reverse the procedure and develop statistical techniques which separate trend, seasonality, and cycles from composite data.

3. How could the least squares regression technique of forecasting, which yields a linear function, be modified to yield curvilinear functions? Discuss the mathematical requirements to achieve such a model; then develop such a mathematical model.

4. How would you analytically assess the trade-off using moving averages between minimizing lag involving few time periods and minimizing the optimal set of weights for n time periods?

5. In a weighted moving average, how would you analytically determine the optimal set of weights for n time periods?

6. Consider the alpha factor in exponential smoothing. How would you determine the appropriate percentage to use? Would the alpha factor be different for different classes of inventory items? Should it remain fixed over time or be changed for the same inventory items? What factors should you consider in arriving at answers to these questions?

7. Compute and plot response patterns to an impulse in demand from Q to $2Q$ with a return to Q in the next time period with the alpha factor set at 0.50, 0.20, and 0.05. Repeat the process assuming a step increase from Q to $2Q$. Which alpha factor provides the best damping response in both the impulse and step increase cases? What factors would you consider in choosing the appropriate alpha factor to use to effect satisfactory impulse or step detection and disequilibrium damping?

Order Quantity and Reorder Point Decision Systems

Determination of Order Quantities

9

In the discussion of the inventory planning and control subsystem it was pointed out that one of the four analytical decision processes involved in the subsystem concerns the determination of the appropriate quantities to be ordered and carried in inventory. In this chapter, several aspects of the order quantity decision will be examined.

The output of any order quantity decision process is an indication of how many units of any particular inventory item should be ordered at one time for replenishment of raw materials inventories or how many units of any particular inventory item should be produced for in-process and finished goods inventories. The decision process used in the determination of order quantities would vary, of course, according to the type of inventory planning and control system in use.

In a time-based or periodic system, the quantity of the order placed to replenish stock at the end of each period (week, month, quarter, etc.) would vary and simply be the difference between the maximum inventory and the stock level projected at the end of each time period. Therefore the focus of attention of the analyst is not placed on the quantity for each period but on the maximum inventory level. The maximum inventory serves as a constraint, as indicated in Figure 7-9, and analysis of usage rates, in conjunction with it, determines the order quantity for any given period of time. The magnitude of safety stock levels also affects the decision concerning the maximum inventory level, since protection against stockouts can be maintained by either raising the safety stock or raising the maximum inventory level. Analysis of these levels is based primarily on the costs of carrying inventories.

In a quantity-based system, such as a two-bin system or one with fixed maximums and minimums, the quantity to be ordered remains relatively fixed as the replenishment cycle is repeated over time. In such a case, the determination of the quantity to be ordered is based on minimizing the costs of carrying inventories and the costs of placing purchase or production orders.

The subsystem designed to determine optimal order quantities is known as the economic order quantity subsystem. Its inputs include two classes of costs, ordering costs and carrying costs. The ordering costs include costs of making requisitions, analysis and selection of vendors, writing purchase orders, following up on the orders, receiving materials, inspecting them, updating inventory records, and carrying out the necessary paper work to complete the purchase transactions.

In some cases, where bidding is involved, additional costs are involved such as the establishment of specifications, requesting bids, evaluating them, selecting the successful bidder, and arranging for inspection equipment to assure that the items ordered conform to the specifications. In order to minimize ordering costs, the fewest possible orders should be placed per year. This requires ordering large quantities every time an order is placed, in most cases.

In the simplest economic order quantity equation, the ordering cost can be determined by dividing the annual requirements (R) by the lot size (Q) to yield the number of orders required per year. The number of orders can then be multiplied by the procurement costs per order (S) to yield a value for ordering cost.

$$\text{Ordering Cost} = \frac{R}{Q}S$$

Given values of 1,000 units required annually, lot sizes checked at 100 unit intervals, and a procurement cost of \$20 per order yields the ordering costs shown in Table 9-1. A cost curve, typical of many ordering cost curves, which is based on the data in Table 9-1, appears in Figure 9-1. As can be seen from the table and figure, the costs decrease rapidly as order quantities increase; however, the rate of decrease slows substantially as the quantity ordered approaches the quantity required annually.

TABLE 9-1

Incremental Inventory Costs

Q Lot Size	$Q/2 \times C$ Carrying Cost	$R/Q \times S$ Procurement Cost	E Total Cost
100	$ 8.00	$200.00	$208.00
200	16.00	100.00	116.00
300	24.00	66.67	90.67
400	32.00	50.00	82.00
500	. 40.00	40.00	80.00
600	48.00	33.33	81.33
700	56.00	28.58	84.58
800	64.00	25.00	89.00
900	72.00	22.22	94.22
1,000	80.00	20.00	100.00

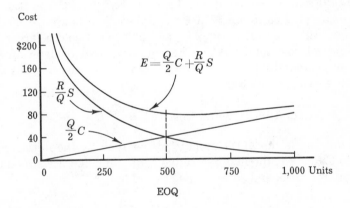

Figure 9-1.   Incremental Inventory Cost Curves

The carrying costs include such items as interest, taxes, obsolescence, deterioration, shrinkage, insurance, storage, handling, and depreciation. In a simple economic order quantity model, these costs are combined to yield one figure which is sometimes expressed as a percentage of the cost of the inventory item in question. In other cases, including the case illustrated here, the carrying cost is expressed in terms of so many dollars and cents per unit per year. The larger the lot size ordered, the larger the average inventory will be and, therefore, the greater the inventory carrying cost per year.

To determine the carrying cost associated with ordering in given quantities, the quantity in question (Q) is divided by 2 to yield the average inventory. This figure is then multiplied by the inventory carrying cost per unit per year (C).

$$\text{Carrying Cost} = \frac{Q}{2} C$$

Given values of 1,000 units required annually, lot sizes checked at 100 unit intervals, and an inventory carrying cost of $0.16 per unit per year yields the carrying costs shown in Table 9-1. The cost curve associated with these carrying costs is linear and is illustrated in Figure 9-1. In the case of carrying costs, the costs increase as the order quantities increase.

To determine the optimal quantity in terms of cost minimization it is necessary to minimize the total incremental cost (E). That quantity will be associated with the minima on the incremental cost curve shown in Figure 9-1 and is known as the economic order quantity (EOQ). From Table 9-1 it can be seen that the EOQ is 500 units where carrying costs are $40.00 and ordering costs are $40.00. Note that quantities larger or smaller than 500 units yield a larger total incremental cost than $80.00. The equation used to determine the total incremental cost is simply the sum of the carrying costs and ordering costs as follows:

$$\text{Total Incremental Cost } (E) = \frac{Q}{2} C + \frac{R}{Q} S$$

To determine the EOQ directly, rather than by the iterative process in Table 9-1, the first derivative of the function E can be taken, set equal to zero, and solved for Q as follows:

$$\frac{dE}{dQ} = \frac{C}{2} - \frac{R}{Q^2} S$$
$$dE/dQ = 0$$
$$Q = \sqrt{\frac{2RS}{C}}$$

Given the values $R = 1,000$, $S = \$20.00$, and $C = \$0.16$, the EOQ is determined to be 500 units as follows:

$$EOQ = \sqrt{\frac{2 \times 1,000 \times 20}{.16}} = 500 \text{ units}$$

This economic order quantity equation is a simple one. Much more elaborate models have been developed over the years by many researchers and industrial practitioners. In determining the appropriate model to use, an analyst should carefully study the cost-quantity relationships of all of the input variables. Only when these relationships reflect the actual conditions under which the inventory system operates will the model be appropriate for decision making purposes in the inventory planning and control subsystem. Since the validity of the cost inputs is directly tied to the validity of the output *EOQ* for decision purposes, further consideration will be given to approaches which might be taken to develop valid cost inputs.

Cost Inputs and the *EOQ*

When evaluating cost inputs for *EOQ* purposes it should be noted that costs derived from accounting records may be misleading. Accounting costs are developed for reporting purposes according to generally accepted accounting principles. Some costs developed for decision making differ from those developed by accounting since the objectives of cost collection necessarily differ from those of decision making. For example, in many accounting records, costs are reported in the aggregate, and include both fixed and variable costs. Since fixed costs cannot be changed by an *EOQ* decision (by definition of a fixed cost) they are disregarded in *EOQ* decisions where the focus is on the variable or incremental costs.

Similarly, sunk costs represent investments resulting from past decisions and are not affected by future decisions. For example, the cost of constructing a warehouse for storage of inventories is a sunk cost which will remain a part of the firm's overhead whether it contains inventory or not. On a continuing basis, the property taxes on a warehouse are fixed costs and continuous whether or not inventories of items are stored in the warehouse. In both cases, although accounting principles might dictate that the costs be attributed to the carrying of inventories, the fact is that a change in policy regarding inventory quantities will not change them. For this reason they are disregarded for decision purposes.

Opportunity costs represent an important component in the analysis of optimal order quantities. They may be applied in a number of ways. The largest opportunity cost usually is associated with the funds tied up in the investment in inventory. Some value should be associated with the lost opportunity to use these funds in other ways in the business. Some firms use the prevailing interest rate on the money invested which certainly constitutes a conservative approach.

Others use an expected rate of return should the money be used in the firm. The rate of return applied is often that experienced by the firm in terms of its return on investment. In a number of companies no attention whatever is given to the opportunity cost of invested capital in inventories which leads to an understatement of carrying costs and consequently larger inventories than should be maintained to minimize total incremental costs.

Another cost factor which bears investigation when cost inputs are determined is the ordering cost and its method of determination. Undoubtedly the easiest way to arrive at such a cost is to gather the costs of operating the purchasing and receiving departments for the past year and divide this by the number of orders processed. The quotient will be the mean ordering cost. Two problems are overlooked when this method is used. First, the costs are historical accounting records and usually contain overhead and other fixed costs which will not change with a change in the EOQ decision. In some cases the mean ordering cost may be taken from even older accounting records and may be totally out of date, particularly if electronic data processing applications have been made in the interim.

The second problem associated with the use of a mean ordering cost is that it is not a marginal cost and the economic order quantity concept is based on marginal analysis. That is, should order activity increase or decrease by 10 per cent, for example, it is doubtful whether the mean ordering cost would change much at all providing it contains overhead and other fixed costs and the change does not require an additional order clerk (or does not result in the dismissal of an order clerk). About the only change in ordering cost in this case would be the change in cost of direct materials — order forms, requisition slips, receiving slips, telephone calls, postage, and the like. These costs may be negligible over a range of plus or minus 10 per cent in ordering activity. Thus the application of a constant factor (S) to the number of orders required per year (R/Q) may be misleading.

In order to overcome this problem some firms use different values for the ordering cost (S) at different levels of operation. Since increases in the number of orders placed per year will result in linear increases in some materials costs and step function increases in others (such as the acquisition of additional personnel to handle larger volumes of orders), different ordering costs will be assigned to different ranges of operations.

The problems of using a constant for the inventory carrying cost (C) are similar to those discussed with reference to the mean ordering cost. Many companies use a constant percentage of the value of inventory items. In practice, however, it is apparent that while the opportunity cost of money and taxes may be prorated as a constant percentage, factors such as storage costs, obsolescence costs, deterioration costs, insurance costs, and others peculiar to particular items may vary widely among different items in inventory.

The differences in treatment of carrying costs can be seen in the results of a 1965 survey of the 500 largest industrial corporations. Of the 292 respondents the average carrying cost reported was 14 per cent, while the range varied from 0 per cent to 30 per cent.[1]

Although the economic order quantity concept represents a useful analytical approach to inventory and materials management, a particular model should be built around the conditions existing in a firm. Those responsible for the decisions concerning inventory and materials management should be thoroughly familiar with the assumptions of the model. Needless to say, most models are oversimplified. However, by building models to fit a particular application, significant savings can be achieved.

Some of the questions which require answers for proper tailoring of the model to a particular firm are as follows:

— With respect to the money tied up in inventory, will the firm use the present cost of money, short-term interest rates, or some opportunity cost associated with alternative uses to which the firm could put the funds?

— With respect to storage costs, will the firm consider sunk or fixed costs associated with warehousing or only the opportunity costs of space occupied by the inventory, and if the latter, how will they be determined?

— With respect to insurance costs, will the firm prorate the insurance premiums or charge a presumed rate if it self-insures its inventories?

— Will taxes be prorated and will consideration be given to reduced tax charges if the firm minimizes inventory levels on the date the taxes are applied to inventories?

— Does the firm experience fad and fashion obsolescence of an unpredictable nature or does it remain fairly constant over time?

— With respect to deterioration, will differential rates be applied to different inventory items or will a prorated value be used?

---

[1] T. M. Whiten, "Report on an Inventory Management Survey," *Production and Inventory Control*, January, 1966, p. 30.

— Will materials handling costs be included or excluded from the model?

— How will the costs of breakage and pilferage be assessed and appropriately allocated?

These questions and many others should serve as sufficient warning to the user of standard *EOQ* formulae that the application of simple models in particular firms can lead to improper decisions. However, once sufficient care has been exercised to develop a model to fit the firm and when the decision makers associated with inventory and materials management understand its assumptions and limitations, it can provide a meaningful improvement over the widespread practice of strictly subjective decision making.

Modifications of the Basic *EOQ* Concept

One of the fundamental misconceptions about the economic order quantity concept is that the minimum cost will provide the only realistic answer to optimizing inventory and materials management practices. Whether the particular value resulting from the application of the economic order quantity equation is of critical importance or not is dependent on the shape of the total incremental cost curve. The shape of the curve is a good indicator of the importance of the *EOQ* determined by the equation. This concept can best be demonstrated by examining two hypothetical situations depicted in Figures 9-2 and 9-3.

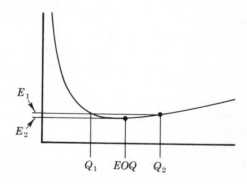

Figure 9-2.   Cost Curve with a Non-critical *EOQ*

In Figure 9-2 a cost curve with a non-critical *EOQ* is depicted. In this case, data similar to that used to generate Figure 9-1 result in

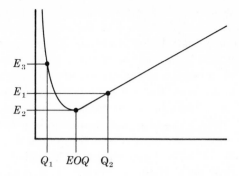

Figure 9-3.   Cost Curve with a Critical EOQ

incremental costs yielding a shallow cost curve. The economic order quantity is at the minima of the curve and yields a cost of E_2. If management were to purchase in lot sizes other than the EOQ, either larger or smaller of the magnitude Q_1 or Q_2, the resultant increase in incremental costs would be negligible, i.e., an increase from E_2 to E_1. In this case, violation of the EOQ decision rule might be quite reasonable to take advantage of other factors of a subjective nature such as the threat of a strike which might shut off the source of supply for some time. Such a factor is not included in standard EOQ formulae but in the case of Figure 9-2 might be hedged against, at moderate cost, by ordering Q_2 units rather than ordering the EOQ.

In Figure 9-3 a cost curve with a critical EOQ is depicted. The same quantity differences from the EOQ are tested. The results in terms of managerial actions are likely to be quite different, however, since the cost consequences are much greater if quantities other than the EOQ are ordered. For example, if Q_2 units are ordered, the higher carrying costs will yield a higher cost (E_1) than is the case in Figure 9-2. Similarly if Q_1 units are ordered, the higher procurement costs will yield a much higher incremental cost than is the case in Figure 9-2. In brief, there is limited latitude for managerial discretion in departing from the EOQ in the case depicted in Figure 9-3 unless a substantial incremental cost increase can be sustained by the firm.

Since the shape of the cost curve is, in one respect, a measure of managerial latitude in departing from the EOQ for decision purposes, it is useful for the analytical system to provide a measure of this latitude. Although acceptable limits could be stated in dollars over the EOQ costs with information outputs generated to yield the upper and lower limits of Q which would conform to these limits, it is possible to get a complete tracing of the curve at reasonable cost through the use of a digital plotter attached to a computer.

Another important modification of the standard economic order quantity concept concerns consideration of the effects of quantity discounts. Thus far only the incremental costs of procuring and carrying inventories have been discussed. To determine whether it is appropriate to take a quantity discount or to buy at the EOQ, it is necessary to examine the cost levels at these quantities. To develop comparative data, the costs of materials must be added to the procurement and carrying costs. If the symbol A is used to represent the cost per unit of the item in question, then the equation which can be used for comparative purposes becomes:

$$\text{Total Cost } (TC) = RA + \frac{Q}{2}C + \frac{R}{Q}S$$

Using this equation, the total cost at the EOQ and the total cost at any other quantity can be determined. Graphically the total cost curve will appear to be discontinuous where the quantity discount applies. Three examples are depicted in Figures 9-4, 9-5, and 9-6.

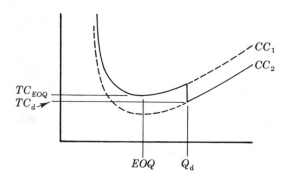

Figure 9-4.   Cost Curve with Favorable Quantity Discount

In Figure 9-4 the cost curve noted as CC_1 depicts the total cost curve which would exist had a quantity discount not been offered. However the quantity discount offered at Q_d is sufficient to reduce the cost of the item A to a point where the total cost curve is lowered to that depicted by curve CC_2. Examination of the solid lines of both curves reveals the total costs for all quantities considering the quantity discount.

To minimize total cost under these circumstances, the quantity discount would be taken and Q_d units ordered rather than the EOQ. This would result in a saving of TC_{EOQ} minus TC_d.

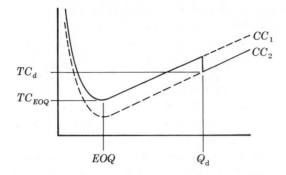

Figure 9-5.   Cost Curve with Unfavorable Quantity
Discount Above EOQ

In Figure 9-5 the cost curve noted as CC_1 again represents the
original total cost curve had a quantity discount not been offered. Cost
curve CC_2 applies to the total cost relationships considering the
quantity discount. As before, the solid lines represent the total cost
relationships which apply to those quantities prior to, including, and
after the quantity discount takes effect.

In this case, in contrast to that in Figure 9-4, the minimum cost is
achieved by not taking the quantity discount. The cost curve reveals
that the high carrying cost at the quantity Q_d, coupled with a negligible reduction in procurement costs and the quantity discount
itself, are insufficient to meet the lower costs available to the firm by
purchasing in smaller quantities at the EOQ. In this case the avoidance of the quantity discount will amount to a sizeable saving
depicted as the difference between TC_d and TC_{EOQ}.

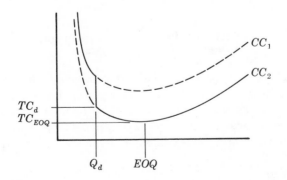

Figure 9-6.   Cost Curve with Unfavorable Quantity Discount
Below EOQ

In both Figures 9-4 and 9-5 the quantity discount volume Q_d was depicted as somewhat greater than the EOQ. In Figure 9-6 the opposite condition is depicted. In this case the EOQ is greater than the quantity discount volume Q_d. Given the set of relationships depicted, it is evident that the quantity discount should not be taken and that the EOQ will yield a lower total cost, even though the EOQ quantity is greater than that at which the quantity discount applies.

The other two possible conditions, Q_d less than EOQ yielding TC_d less than TC_{EOQ}, and TC_d equal to TC_{EOQ}, are not depicted but the approach to analysis of the cost relationships is the same.

Another modification of the basic EOQ model concerns in-process inventories, finished goods inventories, and those few raw materials inventories which do not arrive in single deliveries to completely replenish the stock level. In the case of in-process and finished goods inventories a demand rate exists which tends to reduce inventories but at the same time a production rate exists which tends to increase inventories. Over time these two conditions tend to offset one another and thus alter the basic EOQ relationships. Figure 9-7 depicts the relationships.

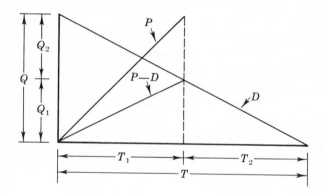

Figure 9-7.   Effects of Differential Production and Demand Rates on Average Inventory Levels

In Figure 9-7 it is assumed that the productive capability of one machine is greater than the demands placed on it for in-process inventory by a succeeding machine. Or, in the case of finished goods inventory management, the productive capability of the assembly department is greater than customer demands for shipment of finished products.

The demand rate indicated in Figure 9-7 can be found by dividing Q by T. In this case demand has a negative slope of 0.5. The production rate is indicated by P and can be determined by dividing Q by T_1 which equals a positive slope of 1.0. Since productive capabilities provide products faster than demand rates can use them, a gradual building of inventory is expected. This will occur over time period T_1 and accumulate to a magnitude of Q_1. The rate of inventory build up will be P-D or a positive slope of 0.5.

To minimize inventories under these conditions the productive capability would be shut down after time period T_1 and the accumulated inventory would be drawn upon during time period T_2 at a rate D. At the end of time period T_2 the cycle would be repeated. The common occurrence of in-process buffer inventories in industry represents the typical case depicted in Figure 9-7.

From the point of view of the EOQ model, there are two modifications which this situation requires. First, it will be noted that under the basic EOQ the carrying cost is determined by multiplying the average inventory $(Q/2)$ by a carrying cost per unit per year (C). In most cases where raw materials inventories are considered and full replenishment of stock is common, the average inventory determined by $Q/2$ would be appropriate. In the case of Figure 9-7 this would be equivalent to a reduction of Q_2 units over the time period T. Replenishment in this case, however, is not instantaneous. It is gradual, at the rate P-D. Therefore the maximum inventory would not be Q but Q_1. Following the assumption of linear usage or demand rates, the average inventory would be Q_1 divided by 2 over the period T.

A more succinct approach would be to reduce the average inventory in the basic EOQ model by the fraction $(P$-$D)/P$ or, in the case of Figure 9-7, by one half. The resultant modification of the incremental cost equation would be as follows:

$$E=\frac{R}{Q}S+\left(\frac{Q}{2}C\right)\left(\frac{P\text{-}D}{P}\right)$$

Similarly the EOQ equation can be modified as follows:

$$EOQ=\sqrt{\left(\frac{2RS}{C}\right)\left(\frac{P}{P\text{-}D}\right)}$$

The results of modification of the basic EOQ equation for non-instantaneous replenishment of stock are 1) a reduction in total incremental costs due to reduced inventory carrying costs (E to E'), and 2) an increase in the quantity indicated as the *EOQ* due to lower inventory levels coupled with the same procurement costs (*EOQ* to *EOQ'*). These relationships are depicted in Figure 9-8.

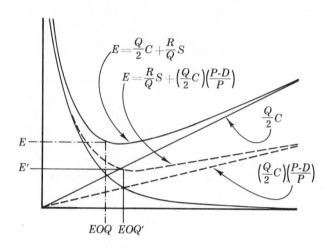

$$E = \frac{Q}{2}C + \frac{R}{Q}S$$

$$E = \frac{R}{Q}S + \left(\frac{Q}{2}C\right)\left(\frac{P\text{-}D}{P}\right)$$

$$\frac{Q}{2}C$$

$$\left(\frac{Q}{2}C\right)\left(\frac{P\text{-}D}{P}\right)$$

EOQ   EOQ'

Figure 9-8.   Effect of Non-instantaneous Replenishment

Determination of Reorder Points

In the prior discussion of inventory planning and control systems, attention was directed to several types of systems. These included 1) the two-bin system, the control loop of which depends on sensing quantities without time or costs; 2) the time-based or periodic system, the control loop of which depends on sensing quantities at periodic points in time; and 3) modifications of both, which serve to dampen system oscillation or amplification of system disequilibrium. These modifications include determination of maximum inventory levels, minimum inventory levels, and safety stocks which serve as constraints on the systems and as sensors to prevent overstocking and/or stockouts.

The fundamental relationships in the determination of reorder points exist among 1) usage rates, 2) lead times, and 3) desired levels of inventory or sensing points such as maximums, minimums, and safety stocks. The measures of effectiveness of any reorder point system which answers the question When should an order be placed? can be measured in terms of the quantities of inventory on hand as the system operates over time subject to endogenous and exogenous variables and in terms of the relative costs of the system.

Figure 9-9 illustrates the basic relationships in a typical reorder point system. Under this construct the reorder point can be determined by multiplying the lead time (LT_i) by the usage rate (Q/T_i) and adding the product to the safety stock (S). The equation is as follows:

$$ROP = \left(\frac{Q}{T_i} \times LT_i\right) + S$$

The reorder point in units then would be the sensing point at which a replenishment order would be placed. If the usage rate continued

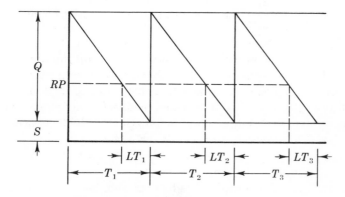

Figure 9-9.   Reorder Point System

to be the same over the period of the lead time, sufficient stock $(Q$ units) would arrive as the inventory decreased to the safety level. The number of units (Q) may be determined using the economic order quantity subsystem or may be subject to the maximum inventory constraint (maximum minus projected inventory level at end of T_i) to protect against underrunning the maximum should the usage rate or the lead time increase.

Before examining the decision system for reorder points, it is necessary to establish some understanding of the effects of lead times (LT_i) relative to usage periods (T_i) under varying conditions. Given the $T_i : LT_i$ relationships existing in Figure 9-9, the system will operate in a state of equilibrium. However, should lead time increase, other things being constant, the reorder point would be raised. Figure 9-10 illustrates these relationships.

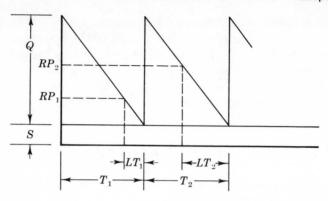

Figure 9-10.   Effect of Lead Time on Reorder Point

In Figure 9-10 the lead time increases from LT_1 to LT_2 with a cor-responding increase in the reorder point. If the information input to the reorder point subsystem is rapid enough during usage period T_2, a new reorder point can be established prior to depletion of the inventory to the level RP_2. Such rapid feedback is necessary to main-tain the system in a state of equilibrium. Should the feedback be too slow, a stockout is likely to occur if the order is placed at RP_1. For that reason LT is subscripted to carry a variety of values as the system is operated over time.

Similarly, the usage rate will affect the reorder point as indicated in Figure 9-11. Assuming the Q, S, LT_1, and LT_2 are constant and that only the usage rate changes, i.e., T_1 decreases to T_2, then the reorder point again is raised from RP_1 to RP_2. Again feedback infor-mation with respect to usage rates (Q/T_i) must be rapid enough for the system to react sufficiently to maintain a state of equilibrium. Otherwise, in the case of Figure 9-11, a stockout may occur before the new stock of inventory arrives.

Another situation affecting the reorder point concerns those cases in which the lead time LT_i exceeds the usage period T_i. When this occurs, the reorder point must instigate replenishment at least one period prior to the period in which the inventory level would reach the safety level. This condition is illustrated in Figure 9-12.

In the case where LT_i exceeds T_i, as in Figure 9-12, the reorder point can be determined quantitatively. For example, if one con-siders usage period T_4, it is apparent that the reorder must occur in the latter part of T_3 to get replenishment by the time the inventory level reaches the safety stock level in T_4. Similarly, the order for usage period T_3 must occur LT_3 days prior to the inventory level reaching S for period T_3.

Using the equation for determination of the reorder point noted earlier, $(Q/T_i \times LT_i) + S$, would yield a reorder point above the maximum inventory level in this case. Thus it must be adjusted to account for the fact that lead time exceeds usage time. The equation can be expressed as follows:

$$ROP = \left(\frac{Q}{T_i}\right) (LT_i - T_i) + S$$

This equation will yield a reorder point sufficient to cover the usage in period T_{n-1} and the usage in period T_n to the point where inventory levels reach the safety stock level.

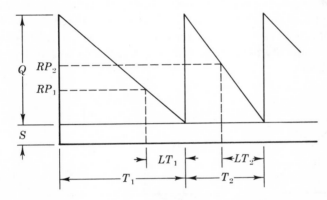

Figure 9-11.   Effect of Usage Rate on Reorder Point

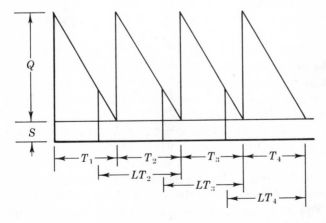

Figure 9-12.   Effect of Lead Time Exceeding Usage Time

Another factor to be considered is the presumed linearity of usage rates. All figures dealing with reorder points depicted thus far have assumed linearity. This, in fact, is not always representative of the physical system, and to the extent that the assumptions of the information and decision system do not reflect the physical system, the usefulness of the equations involved is limited.

To gain a clearer understanding of the assumption of linearity, three cases might be examined. In the first case the product (for example, gasoline in a stationary engine which runs continuously at a constant speed) is used in a fashion which could be depicted as a linear function. In the second case (for example, cleaning solutions pumped through pipelines every other day) the product is used periodically on a batch basis where the inventory is depleted by a constant amount every other day. In the third case (for example, parts kept in stock for repair of equipment) the usage will vary depending on breakdowns, the particular pieces of equipment involved, and the parts required. These three cases are depicted in Figure 9-13.

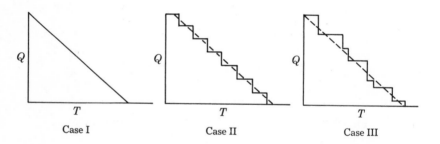

Case I   Case II   Case III

Figure 9-13.   Three Cases of Assumed Linear Usage Rates

In case I the assumed linearity of usage is valid. In case II the asumption of linearity, although not true in fact, represents the status of the physical system over time with sufficient validity to be acceptable. In case III the assumption of linearity is suspect. Anyone who has been associated with spare parts inventory management is aware of the problems of prediction of usage and the typical conditions of stockout or very low turnover which characterize many items in stock.

When confronted with uncertainty with respect to usage rates, deterministic models should be modified to include probabilistic inputs. The same probabilistic considerations also are reflected in circumstances where lead times are subject to variance of a nondeterministic nature. In addition, since the determination of safety stocks is a function of lead times and usage rates, this decision, as well, may rest on probabilistic factors if conditions affecting lead times and usage rates are uncertain.

In conditions of uncertainty a useful analytical aid is a frequency distribution of the occurrences with measures of the mean and standard deviation. In many cases a normal distribution may be applicable; in others a Poisson distribution may be appropriate; yet in still others various degrees of skewness may best reflect the characteristics of the inventory system.

After observing the records of usage rates and lead times, the analyst could determine the probabilities associated with their occurrence. Where such past records are not available the analyst could solicit informed opinions on maximums, minimums, and estimates of the nature of the distribution. He then could set confidence limits with respect to the risk of overage or underage he is willing to accept under conditions of uncertainty and specify the vector of acceptable usage rate variation or lead time variation which could be tolerated. Figure 9-14 depicts a normal distribution of usage rates and Figure 9-15 depicts a skewed distribution of lead times.

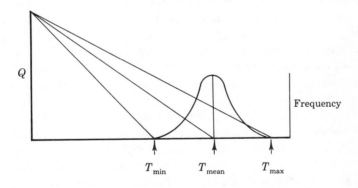

Figure 9-14. Normal Distribution of Usage Rates

Under the conditions depicted in Figures 9-14 and 9-15, one conservative inventory decision process might involve taking the highest usage rate (Q/T_{min}) times the longest lead time (LT_{max}) to develop a reorder point which would insure that the system does not produce stockouts. Although this is a possible solution, it does not solve the problem because the safety stock which would result would be very expensive in terms of carrying cost.

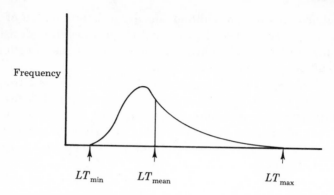

Figure 9-15.   Skewed Distribution of Lead Times

At the finished goods inventory level one might compromise by evaluating desired service rates to customers and comparing the costs of customer ill will due to stockouts against the cost of carrying sufficient safety stock to protect customers from stockouts. Assuming a normal distribution and knowing the mean demand rate and standard deviation, it is possible to compute the safety stock level and reorder point which would provide a stated level of service. This is more reasonable than attempting to maintain 100 per cent service levels as is implied using $ROP = (Q/T_{min} \times LT_{max})$. This area of analysis can be translated into a mathematical model for decision purposes, although the inputs to the model are quite subjective.

The problems of determining accurate carrying costs of inventory have already been explained in connection with the discussion of the economic order quantity. The problems of determining the costs of stockouts affecting customers are even more difficult. One could develop a cost curve of decreasing stockout costs and increasing carrying costs as safety stock levels are increased. By minimizing total incremental cost, one could develop an optimal answer. However, when the cost inputs are ill defined, particularly as is the case with stockouts, it is often more prudent to rely on judgment and experience than on a mathematical representation of the situation which may or may not reflect reality.

At the raw materials inventory level and the in-process inventory level, such a model may be more realistic since costs of stockouts may be more accurately determined. The factors which might be considered include costs of down-time on the machines involved, lost labor productivity, cost of rush orders, expediting, special handling, air freight, higher prices for small rush orders, and the disruption of production schedules. Although these costs may be more accurately defined than the stockout costs associated with lost sales to customers, there are still many factors for which accurate costs are not available from the firm's cost accounting records.

In those cases where the frequency distributions do not conform to standard statistical distributions, the analyst might use the Monte Carlo method of simulation to determine the expected values for irregularly distributed usage rates, lead times, and, consequently, safety stocks. By repeating the simulations many times, a normal distribution of results may be generated. This distribution can then be utilized to develop confidence levels for purposes of estimating appropriate levels of safety stock to maintain desired service rates.

Monte Carlo Simulation of Inventory Fluctuations

The Monte Carlo technique as applied to the simulation of inventory fluctuations operating with particular decision rules can be demonstrated in the following example. Assume that a certain inventory item is required on a daily basis and that its usage rate is random. From past records it is known that the following distribution represents the probabilities of usage:

Daily Requirements (Units)	Probabilities	Random Number Ranges
5	.10	0–9
6	.10	10–19
7	.20	20–39
8	.30	40–69
9	.20	70–89
10	.10	90–99

Similarly, when a decision is made to reorder, it is known from past records that the number of delivery days required varies randomly and that the range of delivery days required and their probabilities are as follows:

Delivery Time (Days)	Probabilities	Random Number Ranges
10	.05	0–4
11	.10	5–14
12	.30	15–44
13	.20	45–64
14	.15	65–79
15	.10	80–89
16	.05	90–94
17	.05	95–99

Assume that the number of units in inventory at the beginning of the simulation is 100. If an out-of-stock status occurs, then the back orders are immediately filled when a replenishment stock arrives. The inventory system is designed around the reorder point concept and a decision rule is to be tested for stockout and overstock characteristics. The decision rule to be tested is that a replenishment order for 100 units will be placed whenever the inventory at the end of the day falls below 100 units.

To simulate the behavior of the inventory system in terms of aggregate input-output fluctuations, it is necessary to generate random numbers which, in turn, will reflect the probabilities of usage and delivery time when applied against the random number ranges cited above. As a test, 100 days of experience are to be simulated. The simulation appears in Table 9-2.

TABLE 9-2

Day	Usage Random No.	Daily Require- ments	End of Day Inventory	Units Received	Delivery Random No.	Delivery Time in Days
0			100			
1	18	6	94		46	13
2	32	7	87			
3	14	6	81			
4	26	7	74			
5	77	9	65			
6	93	10	55			
7	06	5	50			
8	42	8	42			
9	15	6	36			
10	38	7	29			
11	53	8	21			
12	19	6	15			
13	06	5	110	100		
14	79	9	101			
15	95	10	91		92	16
16	12	6	85			
17	26	7	78			
18	82	9	69			
19	38	7	62			
20	70	9	53			
21	76	9	44			
22	06	5	39			
23	38	7	32			
24	77	9	23			
25	96	10	13			

TABLE 9-2 (cont.)

Day	Usage Random No.	Daily Require- ments	End of Day Inventory	Units Received	Delivery Random No.	Delivery Time in Days
26	62	8	5			
27	08	5	0			
28	73	9	(−9)			
29	88	9	82	100	27	12
30	37	7	75			
31	14	6	69			
32	82	9	60			
33	63	8	52			
34	27	7	45			
35	84	9	36			
36	92	10	26			
37	02	5	21			
38	71	9	12			
39	63	8	4			
40	50	8	96	100	38	12
41	17	6	90			
42	08	5	85			
43	38	7	78			
44	89	9	69			
45	94	10	59			
46	03	5	54			
47	27	7	47			
48	38	7	40			
49	88	9	31			
50	17	6	25			
51	07	5	120	100		
52	22	7	113			
53	83	9	104			
54	17	6	98		63	13
55	42	8	90			
56	26	7	83			
57	18	6	77			
58	63	8	69			
59	88	9	60			
60	92	10	50			
61	02	5	45			
62	28	7	38			
63	37	7	31			
64	16	6	25			
65	44	8	17			
66	83	9	108	100		
67	99	10	98		16	12
68	28	7	91			
69	02	5	86			
70	73	9	77			

TABLE 9-2 (cont.)

Day	Usage Random No.	Daily Require- ments	End of Day Inventory	Units Received	Delivery Random No.	Delivery Time in Days
71	46	8	69			
72	15	6	63			
73	22	7	56			
74	84	9	47			
75	77	9	38			
76	92	10	28			
77	17	6	22			
78	26	7	115	100		
79	14	6	109			
80	38	7	102			
81	39	7	95		76	14
82	55	8	87			
83	32	7	80			
84	27	7	73			
85	16	6	67			
86	04	5	62			
87	36	7	55			
88	52	8	47			
89	19	6	41			
90	26	7	34			
91	83	9	25			
92	95	10	15			
93	77	9	6			
94	63	8	98	100	55	13
95	26	7	91			
96	01	5	86			
97	46	8	78			
98	42	8	70			
90	73	9	61			
100	21	7	54			

Although 100 iterations represents a very limited simulation of the behavior of the inventory system, it indicates that the decision rule of ordering 100 units when the end-of-day inventory drops below 100 units is a good one considering the probability distributions related to usage rates and delivery times. During the simulation above, only one stockout occurs of nine units and lasts but one day. The average ending inventory for each iteration is slightly less than 15 units. By increasing the reorder quantity it would be possible to eliminate the stockouts, however, this would result in somewhat higher levels of ending inventories for each iteration. To test such a

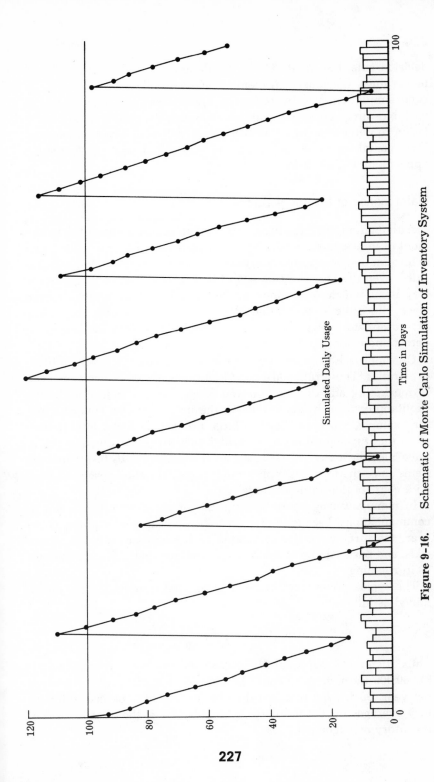

Figure 9-16. Schematic of Monte Carlo Simulation of Inventory System

Time in Days

Simulated Daily Usage

227

decision rule, the Monte Carlo simulation could be repeated using new values. Obviously an element of trial and error is present in such simulation, but by assigning appropriate cost factors to the variables involved, it is possible to measure the results of the simulation to determine which decision rule will tend to minimize total incremental costs. A schematic of the simulation in Table 9-2 appears in Figure 9-16.

Maintenance of Inventory Turnover

Although an effective inventory system requires careful development of forecasts, determination of appropriate order quantities, and determination of appropriate times to place orders for those quantities, even the best of systems will be characterized by some inventory items which, over time, can be classed as slow-moving or dead stock. Since the accumulation of such inventory items is inevitable, it is useful to incorporate control capabilities within the system to detect and correct this phenomenon.

One approach to control over inventory turnover is the use of turnover ratios. These ratios are most highly developed in firms engaged in marketing and distribution and often serve as indicators of the profitability of operations. Indeed, the inventory of a supermarket, for example, represents such a large proportion of the total investment that turnover must be very high to insure even a modest profit. Slow moving items are replaced on the shelves with faster moving items since the former simply cannot justify the shelf space required on a contribution to profit basis.

In manufacturing operations the use of turnover ratios is not as common, in part because it often is not as critical as in retailing. However, the turnover ratios associated with storeroom supplies, materials, and parts can be developed and used in much the same manner as in retailing.

The typical inventory turnover ratio used in marketing is determined as follows:

$$\text{Turnover Ratio} = \frac{\text{Annual Sales}}{\text{Average Inventory}}$$

Thus, if annual sales equal $10,000,000 and average inventory equals $1,000,000 the inventory could be viewed as turning over ten times per year. Such a figure might then be contrasted to published inventory averages for the industry as an indicator of the effectiveness of inventory management.

Although this figure represents an aggregate approximation of inventory turnover, it can be refined in scope to focus attention on turnover of a particular class or category of inventory items. Similarly, the annual time period can be reduced to provide more rapid feedback of information.

In the operations area the same type of ratio can be applied in those instances where finished products are produced to stock. In such cases the annual sales of any stock item can be divided by the average inventory of that item to yield a turnover figure. These data are available from the historical records maintained by the inventory stock status subsystem.

Once these ratios are developed for all stock items, the turnover figures can be compared among items. This comparison often reveals those stock items which are not moving into the distribution system rapidly and which, therefore, represent excessive funds tied up in finished goods inventory. Subsequent evaluation of slow moving items with very low turnover ratios may lead to efforts to accelerate sales (increase the numerator of the ratio), reduce the average inventory in stock (decrease the denominator), or both.

A similar approach can be taken with inventories of supplies, parts, and tools used in operations. The ratio would be modified somewhat as follows:

$$\text{Turnover Ratio} = \frac{\text{Usage in Time Period } T}{\text{Average Inventory in Time Period } T}$$

This ratio represents how many times the inventory of particular supplies, parts, or tools is used in a particular time period such as a month, quarter, or year. A comparison of such ratios may indicate the slow turnover items which represent money tied up in inventory and may be released for more profitable uses.

The opportunity cost of money represents one of the costs of carrying excessive inventories. To it can be added other costs such as taxes, risks of obsolescence, deterioration, shrinkage, insurance, storage, handling, and depreciation.

In total, these costs often justify the necessary means to keep inventory levels relatively low and turnover ratios high. However, such measures incur certain risks not apparent in inventory management at the wholesale or retail levels in the materials management system beyond the manufacturing stage. The risk of not having available replacement parts for equipment, for example, can mean excessive machine down-time, lost production, idle manpower, and expensive maintenance which may more than offset the savings achieved through reducing inventories. Similarly, the lack of tools and certain supply items can lead to the same results. Any consideration of target inventory turnover levels should include an analysis of these offsetting costs.

Since there is an element of uncertainty with respect to the usage rates of supplies, parts, and tools, especially with respect to breakdowns, it is useful to apply probabilities to the events to ascertain the appropriate inventory policy. As an example of the application of probabilities, consider the following situation. One department utilizes eight identical machines and the probability distribution associated with breakdowns is Poisson, determined as follows:

$$P_x = \frac{\mu^x e^{-\mu}}{x!}$$

where x = number out of order, μ = average number out of order, and $e = 2.7183$.

The production loss associated with each machine out of order is $100 per day. The question to be resolved is how many replacement parts should be kept in the parts inventory to minimize total costs. A set of parts for one machine is assumed to cost $70 including inventory carrying costs. If no sets of parts are provided, the expected production loss is calculated at $199.10, as follows:

Number of Machines Out of Order	Probability	Production Loss	Expected Loss
0	0.135	$ 0	$ 0.00
1	0.270	100	27.00
2	0.270	200	54.00
3	0.180	300	54.00
4	0.090	400	36.00
5	0.036	500	18.00
6	0.012	600	7.20
7	0.003	700	2.10
8	0.001	800	0.80
			$199.10

If one set of replacement parts is in the parts inventory, then the probability of loss due to breakdowns is reduced and the expected production loss becomes $113.70, as in the following table.

Number of Machines Out of Order	Probability	Production Loss	Expected Loss
0	0.135	$ 0	$ 0.00
1	0.270	0	0.00
2	0.270	100	27.00
3	0.180	200	36.00
4	0.090	300	27.00
5	0.036	400	15.20
6	0.012	500	6.00
7	0.003	600	1.80
8	0.001	700	0.70
			$113.70

If two sets of replacement parts are in the parts inventory, then the probability of loss due to breakdowns is further reduced and the expected loss becomes $53.70, as follows:

Number of Machines Out of Order	Probability	Production Loss	Expected Loss
0	0.135	$ 0	$ 0.00
1	0.270	0	0.00
2	0.270	0	0.00
3	0.180	100	18.00
4	0.090	200	18.00
5	0.036	300	10.80
6	0.012	400	4.80
7	0.003	500	1.50
8	0.001	600	0.60
			$ 53.70

Of course, as extra sets of replacement parts are kept in inventory, the expected production losses would continue to diminish. By minimizing these costs, an inventory decision could be made to stock one set of replacement parts (assuming replenishment of stock in one day from the supplier) as follows:

Expected Loss of X Machines Being Out of Order	Cost of Replacement Parts	Total Cost
$199.10	$ 0.00	$199.10
113.70	70.00	183.70
53.70	140.00	193.70

Even though the probabilities of expected production loss will continue to decrease in a manner approximating a negative curvilinear function, the positive linear function associated with replacement parts $(Y = 0 + \$70\ X)$ nullifies its effect for consideration beyond three sets of parts.

In other situations the use of standard statistical distributions may not be appropriate if past history indicates that usage patterns are highly irregular. In such cases, Monte Carlo simulation may provide a useful prediction of usage requirements and, therefore, inventory policy. An example will clarify this approach.

Consider that a particular type of cutting tool is used interchangeably on three machines, A, B, and C. The clerk responsible for the tool room checks his records and finds that the type of tool in question wears out in varying periods of time due to the wide variety of operations performed, materials used, and workmen assigned to particular jobs. His records indicate the last 100 tools issued and returned to the tool room lasted the following number of days:

Days	Number of Tools	Range
1	5	00–04
2	20	05–24
3	5	25–29
4	10	30–39
5	10	40–49
6	40	50–89
7	10	90–99
	100	

By assigning ranges of two-digit numbers which correspond to the tool usage history, the irregular distribution is represented. In the Monte Carlo technique a random number generator is used to predict what is likely to occur. If it is assumed that a thirty-day supply is to be ordered, because of high procurement costs, then the problem becomes one of determining how many tools should be ordered to supply the three machines for this period of time.

Let us assume that the first random number is 17. Then the expected usage on Machine A would be 2 days, since this number falls in the range corresponding to 2 days. The next random number might be 33. Machine B therefore would be assigned 4 days. If the next random number is 52 then Machine C would be assigned 6 days. The next series of random numbers would again cycle through Machines A, B, and C, until the thirty-day requirement was met on all three machines. The random numbers and tool usage forecasts follow:

Machine	Random No.	Days	Total Days
A	17	2	2
B	33	4	4
C	52	6	6
A	28	3	5
B	63	6	10
C	17	2	8
A	70	6	11
B	93	7	17
C	31	4	12
A	58	6	17
B	73	6	23
C	82	6	18
A	14	2	19
B	44	5	28
C	91	7	25
A	36	4	23
B	61	6	34
C	08	2	27
A	98	7	30
B	—	—	34
C	40	5	32

The number of tools required from this iteration of the simulation is 20. Intuitively, one is left with an uneasy feeling about the result, however. Therefore the number of iterations is repeated many times, preferably with the use of a computer with an internal random number generator. After many such iterations are performed, the answers, when plotted, will tend to form a frequency distribution which approximates a normal distribution.

Once the normal distribution is developed, the mean and standard deviations can be computed. In order to maintain a particular confidence level in the answer, the user of the technique would order the mean number of tools plus the number of tools times the standard deviation appropriate to the selected level of confidence. Obviously, the answer must be rounded to the nearest integer for purposes of inventory replenishment.

An alternative approach in this example would be the use of a weighted average. An expected tool usage of 4.60 days can be derived by multiplying the percentages of tool usage times their respective usage histories and summing the weighted values as follows:

Days	Percentage	Weighted Days
1	.05	0.05
2	.20	0.40
3	.05	0.15
4	.10	0.40
5	.10	0.50
6	.40	2.40
7	.10	0.70
		4.60

Since three machines require 30 days supply, or coverage for 90 machine days, then the number of tools required would be 90 machine days divided by 4.6 days per tool, or 20 tools rounded to the nearest tool. The use of either approach would be appropriate in those situations where the frequency distribution with respect to the variables involved does not conform to a standard statistical distribution.

An alternative approach to the maintenance of inventory turnover involves aging inventories over time. In a manner similar to that used by credit managers to age accounts receivable for collection control, the inventory system can be supplemented with an aging routine to minimize dead stock in the system. Aging reports can be generated at various points in time and for different aging periods. For example, aging reports may list those inventory items for which no usage or demand is reported for 30 days, 60 days, and 90 days. Weekly review of such listings in disposition meetings should lead to decisions with respect to rework, reallocation, or scrapping of the items.

Theoretically, an inventory system which is replenished by a triggering system for orders designed around reorder points should result in no aged inventory items on order at a given point in time. Should stock levels be reduced at a given point in time due to usage, the item no longer appears on the aging report of inactive inventory items. Thus both the reorder routine and aging routine should be linked within the system.

Periodic reporting of aged inventory items on a weekly basis results in an exception report which continues to draw management attention, week after week, to those items which should be worked out of inventory in one way or another. If coupled with weekly disposition meetings and summary reporting of total dollars tied up in aged inventory, these reports should result in adequate clearing of dead stock.

Inventory aging on a weekly reporting cycle, coupled with disposition meetings, is a time-consuming task for managers. Thus it is important to consider that only certain inventory items be considered for aging reporting. The A-B-C analysis of value and usage of inventory items discussed earlier is applicable in this instance. The high value items and/or those which constitute a large proportion of inventory, even though they are low value individually, are prime candidates for aging on a routine basis. However, most of the inventory items should be subjected to an aging report semi-annually to clear out items of small value which would otherwise go unnoticed year after year.

Summary of the Inventory Planning and Control Subsystem

The inventory planning and control subsystem is one of the four basic subsystems in the operation of the inventory and materials management subsystem. It serves as the analytical nerve center of the larger subsystem and provides answers to several key questions. The answers often result from the application of decision rules objectively determined, although, in some cases subjective overlays are required for realistic results.

The subsystem can be viewed as dynamic in that an information loop operates over time with varying degrees of periodicity depending on the particular planning or control function being executed. The primary information process of the subsystem involves forecasting of inventory requirements. An initial sorting process using A-B-C analysis determines the level of sophistication of the forecasting model which is applicable. Following this sort, a variety of forecasting models can be considered for the determination of requirements, including the following: least squares regression to interpolate future requirements from past sales or usage data; moving averages; exponential smoothing; and variations on these for forecasting seasonal and cyclical demand in addition to trend forecasting.

The nature of these forecasting models is such that they continue to be employed over time as the state of the inventory system changes. This requires an information feedback loop from the inventory stock status subsystem to indicate changes in usage or demand rates. Although this information loop, composed of usage history and forecasting models operating over time, can be developed as a closed loop similar to mathematical models applied in simulation of hydraulic systems, it is important to provide for inclusion of subjective factors which represent non-programmable stimuli from endogenous and exogenous sources.

The fundamental output of the forecasting models is the level of expected requirements for each of the inventory items to which the model is applied. This information is required for long-range planning and also as an input to the next important analytical function in the inventory planning and control subsystem, the determination of how much to order.

The determination of order quantities is based on the forecasted requirements plus a number of other cost factors broadly classed as procurement costs, carrying costs, fluctuations costs, and opportunity costs. Again mathematical models can be applied to individual inventory items. These models will yield the appropriate order quantity to minimize total incremental costs. Once the order quantities have been determined, this information can be fed to the next important analytical function of the subsystem, the determination of when to place the order.

The determination of reorder points results from the application of decision rules based on the quantities developed from *EOQ* analysis plus expected usage. Useful models must provide for the uncertainty which is characteristic of usage rates and delivery times. It is in this area that statistics and Monte Carlo simulation may provide answers concerning the operation of the system over time where the behavior of the system cannot be replicated deterministically.

If one views the inventory planning and control subsystem as starting from an initial non-operative state, then the application of the forecasting models, *EOQ* models, and reorder point models could be viewed as the initial *planning* devices yielding those decisions which would initiate action within the system. Once the system begins to operate, the order quantities are transmitted to the purchasing subsystem at those points in time indicated by the reorder point models. Initial filling of inventories then follows, being tracked through the logistics and stock status subsystem.

The inventory system, however, is composed of more than the inputs to stock via the purchasing and logistics subsystems. The outputs from inventory must also be tracked via the inventory stock status and logistics subsystems. This information is then fed back to the inventory planning and control subsystem and serves to modify the forecasts, *EOQ*'s, and reorder points. Thus these models can be viewed as *control* devices which provide the system with homeostatic response characteristics.

Two subsidiary analytical functions are also performed by the inventory planning and control subsystem. One of these concerns the maintenance of inventory turnover with control over slow-moving items and dead stock. The other is concerned with the related question of disposition of obsolete inventory items. Although a well-designed inventory system will have minimal accumulation of such stock, it nevertheless must be cleared periodically. Thus, information feedback loops should be developed to cover this contingency. Two such feedback loops can be built around 1) turnover ratios as managerial sensors of performance in this area, and 2) aging reports which identify particular stock items as potential dead stock which should be considered for disposition.

Conceptual Problems

1. Consider the problem of ascertaining procurement costs and inventory carrying costs. Using a firm with which you are familiar, list the pertinent factors which should be investigated in arriving at these cost figures. What criteria underlie your selection process?

2. How would you solve cost allocation problems in determining inventory carrying costs and procurement costs? As a case in point, consider the division of fixed and variable costs associated with procurement.

3. With respect to money tied up in inventory, should a firm use the present costs of money, short-term interest rates, or some opportunity cost associated with alternative uses to which the firm could put the funds? What factors influence this decision?

4. When a firm internally finances inventories, there is an opportunity cost of not being able to use these funds elsewhere. Since there are great varieties of alternative uses of funds, how would you determine the opportunity cost?

5. Cost curves with critical and non-critical *EOQ*'s are discussed in this chapter. Where would you set the limits which would indicate critical levels of cost changes? What analytical process would you use to determine these levels?

6. Discuss your reactions as a customer to a store which does not have what you want in stock. Will you wait for the store to order it? Do you take your business elsewhere? If you were the manager of the

store, how would you determine the losses, both present and future, which result from inadequate inventories?

7. One modification of the basic *EOQ* model involves quantity discounts. This factor and changes in market prices in general can impact substantially the order quantity decision. How would you go about forecasting price changes of particular items? How would these forecasts affect your inventory decisions?

8. In the determination of reorder points, a safety stock is usually incorporated as a buffer for the decision making process. Since large safety stocks are costly, improvements in the decision making process can be a real cost saving effort by reducing this buffer. Probabilistic models provide one means of effecting such a reduction. How could they be applied to solving the *EOQ* problems characterized by variations in demand or usage rates, variations in production or inventory rates, variations in lead times for placing orders, or variations caused by probable future price changes?

9. Consider potential applications of the Monte Carlo method in the analysis of inventory systems in an industry with which you are familiar. How would you develop the appropriate set of probabilities to utilize in the simulation procedure?

The Logistics Subsystem

The Logistics Subsystem

The Logistics Subsystem

10

The materials flow network has been discussed up to this point with emphasis on the inventory planning and control subsystem. A closely related subsystem pertains to the rate of flow of materials through the system. Tracking of rate of flow occurs within the logistics subsystem.

As indicated at an earlier point, effective management of materials flow networks requires not only analysis of the capacity of the system and the quantities at critical points, but also the rate of flow through the system at those points between decoupling inventories. In fact, as is illustrated with the mechanical analog, the effective operation of such a flow network reveals that effectiveness is a function of all three factors, capacity, volume at given points, and rates of flow between these points.

It is the function of the logistics subsystem to provide information with respect to rates of flow and also to provide analysis of particular aspects of spatial flow pattern decisions. The critical points involved are shown in Figure 10-1 as valves which provide transit information. The first area where flow rate reporting is critical is between the vendors and receiving. Although absolute quantities in this area are tracked by the inventory stock status subsystem, it is as important to know where the items are while they are in transit and how fast they are moving.

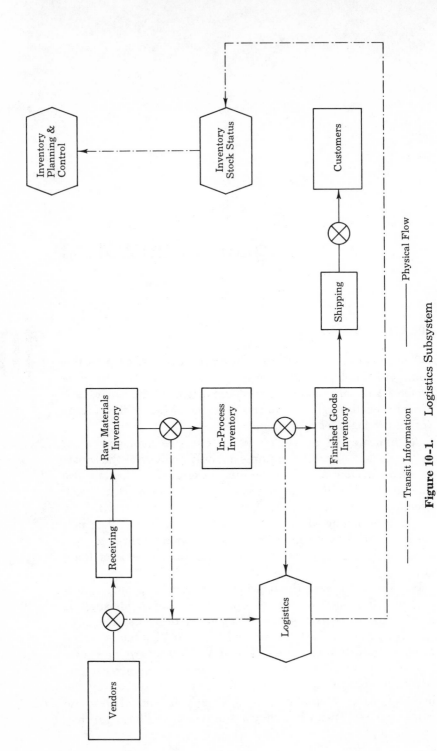

Figure 10-1. Logistics Subsystem

— Physical Flow

——— Transit Information

In some industrial situations it is common to have materials moving to the firm from vendors on different carriers over different routes at different rates of speed. Consider the case of a manufacturing plant which may receive the same raw materials from a number of suppliers, some of which are near the plant and some of which are distant. In addition, the mode of transport may vary so that certain of these raw material items are shipped by truck, rail, and ship. In such a case, the close suppliers may use trucks, those farther away with good rail connections may use freight cars, and those far distant may be able to compete only because of the low cost per ton-mile of ship transport.

The firm must know in each case how much of the item in question is on each carrier. These data are collected in the stock status subsystem. However, this knowledge is insufficient for effective planning and control. The firm must also know the departure time of each carrier and its estimated arrival time, from which rates of flow can be determined. Although this condition of complexity may be viewed as a severe problem to the firm, it actually can be turned to advantage. By having a substantial input stream of raw materials flowing to the firm on different carriers at different rates of flow, the vendors thereby carry the inventory for the firm. The suppliers have tied up capital, and the carriers hold the risk and provide temporary storage of inventory items committed to the firm. If appropriate logistical analysis and tracking takes place, this situation can be turned to an advantage for the firm. Without it, the management of such a diversified input stream can be quite complex.

The second critical point in the firm where logistical data must be acquired and analyzed occurs between raw materials and in-process inventories. The stock status subsystem tracks absolute amounts at the end points mentioned; however, it is also important to assess the rate of flow of raw materials into manufacturing processes where they are transformed and tallied as in-process inventory items.

A simple tally of items might indicate that at the beginning of the day the raw material area had 1,000 units of inventory item A and the in-process inventory area had 2,000 units of inventory item A. At the end of the day the figures might be just the same and a general conclusion might be that nothing happened with respect to inventory item A during the day. In fact, it is possible that 500 units of item A were received and issued from the raw materials inventory area and that the same 500 units passed through the in-process inventory area leaving no net change for the day. It is to overcome this type of misjudgment that the logistics subsystem is inserted to track rate of flow.

Tracking of absolute amounts alone is insufficient for effective closed loop control.

A similar condition exists with respect to the point between in-process inventory and finished goods inventory. Here again there are a number of complex flow patterns. To accurately monitor the status of the system, it is necessary to track not only quantities but also rates of flow between critical connective points. These critical connective points, of course, are much more numerous than those abstract three points used in this discussion — raw materials, in-process, and finished goods inventory. For example, raw material repositories may exist at different points in the plant depending on the nature of the material stored and the physical location of those operations in which it typically is used. In-process inventories may exist as buffer stocks in a number of locations, usually between sequential machine processes with different throughput rates. In some job-lot cases, these inventories may even be moved to special in-process areas where they are simply held until a sufficient order quantity accumulates to justify an expensive set-up on a particular machine. At the finished goods inventory point there may be several finished goods repositories. Some may be near a shipping dock for rail shipment. Others may be near shipping docks designed for trucks. Still others may be warehouse storage points where the product may remain for years, as is the case with stocks of replacement parts which the manufacturer produces to satisfy future requirements of customers who may need them.

A third critical point for the monitoring of rates of flow exists between shipping and the customers. In a systemic sense, if this connective involves all channels of distribution between the firm's shipping dock and the ultimate consumer, it can represent an extremely complex dispersion when viewed as a flow network. Rates of flow may need to be monitored between the firm and fifty wholesalers and, beyond that, between the fifty wholesalers and five thousand retailers. Design problems associated with monitoring systems for such a degree of dispersion fairly boggle the imagination.

At this point we shall examine in an abstract sense the nature of the logistics problem and some of its challenges. As a frame of reference consider the flow network in Figure 10-2.

In this case it can be seen that a number of vendors (V) channel items to raw materials inventory (RM) which then diverge to several in-process inventories (IP) and converge at finished goods inventory (FG) and finally diverge again to the customers (C). It is this problem of divergence and convergence which is at the core of the logistics problem.

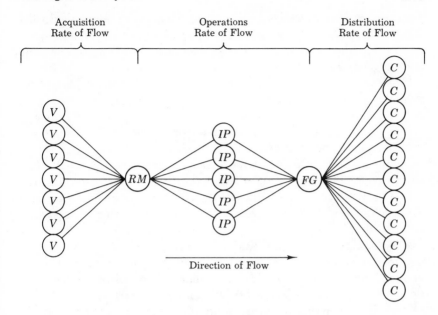

Figure 10-2. Convergence-Divergence Patterns

The objective of a logistics effort is to supply the items required at the "right" places in the "right" quantities at the "right" times at a low cost. Although this is easily stated, it is difficult to specify in more precise terms. In order to specify the right quantities, data are generated from the inventory planning and control subsystem to the purchasing subsystem to the vendors for raw materials. These quantities are determined from the analytical models discussed in Chapter 9 and are based upon considerations of requirements, procurement costs, carrying costs, and reorder points.

The determination of the "right" quantity at each of the in-process inventory points is based on analysis of production scheduling models which consider machine-loading and man-loading problems, plant capacities and current manufacturing requirements. Such problems in themselves can be quite complex.

The "right" quantity at the customer level is dependent on expected demand in some cases, actual orders in others. These quantities, of course, vary from one customer to the next and are further complicated by quantity determinations at the buffer points in the channel of distribution, i.e., wholesalers and retailers.

The "right" place is fairly easy to determine in the case of raw materials and finished goods inventories since there are limited and established respositories for these items. However, the "right" place with respect to in-process inventories and customer inventories presents a more complex problem since these locations change from day to day and, in the case of customer inventories, may represent wide geographic dispersion.

The consideration of the "right" times at which items should be available also requires detailed analysis of the interacting flow patterns and flow rates which converge or diverge. In the broadest sense, one could conceive of an acquisition rate of flow from all vendors to the raw materials inventory, an operations rate of flow from raw materials through the manufacturing process to the finished goods inventory, and a distribution rate of flow from finished goods to the various customers. In an effective system design, these three major flow networks would be balanced in terms of aggregate rates of flow.

A system design which would solve this problem in a very simplified model would be characterized as a continuous process with one vendor, one raw material repository, one in-process inventory location, one finished goods repository, and one customer. In such a system the maintenance of a standard rate of flow which would satisfy customer demand could be easily established at, say, 100 units per hour between each connective point in the system. If the rates of flow began to vary between these connective points, the results would be disequilibrium in the system characterized by the build-up or depletion of particular inventories.

It is unfortunate that a simple system such as that just described does not exist for most firms. Yet the same balance of flow rates does exist in somewhat more complex configurations. Consider for example the oil refinery with its flow rate balancing between raw materials and finished goods. In this case large storage tanks serve as buffers at the beginning and end of the process yet, in between, the flow patterns are carefully balanced and operate continuously. Crude oil enters the system at a particular rate and then it follows particular pipeline routes through a variety of processes to emerge as gasoline, kerosene, diesel oil, lubricants, and other petroleum products on a continuous flow process basis.

To enlarge the concept even further, consider the production and distribution of electricity. At the vendor level, a repository of energy source material is available — a river or waterfall in a hydroelectric station, an atomic energy source or steam source in other cases. This source material is then transformed into electricity and distributed to customers as they demand it. Certain circuits and switching stations assist in the distribution process yet there is no position at which the "product" can be stored. In a sense, as customer demand increases at any one point in the system, the change in flow rate, along with all other customer flow rate changes, immediately feeds back to the production facility which alters its flow rate to match the customer flow rate. Such a system emphasizes how the matching of flow rates in the acquisition phase, operations phase, and distribution phase is essential for smooth operation of the entire system. It also illustrates that such a systemic construct need not be limited to trivial systems with very limited patterns of convergence and divergence between connective points.

As has been pointed out earlier, the tracking and rapid reporting of changes in flow rates is essential for the maintenance of system equilibrium. Without it, a small change near the output stage of the system can feed back to prior stages and cause amplification of the signal — in some cases, the opposite response pattern to that intended. This is the common condition of delayed response on a pattern characterized by oscillation.

Analytical Techniques in the Logistics Subsystem

The analytical techniques, which are applicable in the area of logistics, fall generally into two areas, those associated with the study of rates of flow between connective points and among successive sequential points and those associated with the problems of convergence and divergence. In some cases, models have been developed which attack both problems simultaneously. Model building based on queuing theory has led to useful insights in systems design and systems analysis with respect to problems of rates of flow among connective points. Model building based on the transportation problem associated with linear programming has assisted in the development of effective decision rules in the problem areas of convergence and divergence. Model building based on the Monte Carlo technique of simulation has provided a method of evaluating decision rules where resource allocation problems bear on differential flow rates.

Queuing Theory[1]

Many logistics problems involve the analysis of waiting lines or queues. Questions such as the following are related to queuing problems. How many shipping docks and receiving docks should be

---

[1]Adapted from R. J. Hopeman, *Production: Concepts, Analysis, Control* (Columbus, Ohio: Charles E. Merrill Books, Inc., 1965), pp. 149-161.

built for incoming and outgoing shipments? How many lift trucks should be provided to move products? How many tool crib attendants should be provided to supply tools to workers? How should machines be arranged to minimize waiting lines of products in process? How should traffic patterns be designed inside and outside the plant to minimize waiting lines?

The typical queuing problem involves a bottleneck of some sort. Several aspects of the bottleneck operation can be analyzed by using queuing theory. To understand these aspects, certain terms must be defined in a general way. As a frame of reference for the definitions, we shall assume that we are faced with the problem of determining the number of loading and unloading docks we shall need in the shipping and receiving department.

A *customer* is a person or thing which requires service. In our example, the trucks which arrive at the plant are customers. A *service* is the action wanted by the customer. For the trucks, the service will be loading and/or unloading. A *service facility*, sometimes called a *station* or *channel*, is the people and equipment necessary to provide the service. In the shipping and receiving department the service facility is the dock, materials-handling equipment, and shipping and receiving personnel. A *queue discipline* is the manner in which customers are serviced. In the case of trucks, we shall assume that they are serviced on a first-come, first-served queue discipline. An *arrival distribution* is the way in which customers arrive to form the waiting line or queue. The trucks may arrive at random times. If this is the case, the problem is said to have a *random* arrival distribution. A *service-time distribution* is the distribution of time required to serve the customer. This may take many forms. It may be constant if it takes the same time every time. This may be the case if the trucks carry the same quantity and type of material. In some cases, service-time distributions vary and must be simulated.

All queuing applications have certain things in common. Customers arrive at some rate which has a particular distribution. The customers then may have to join a waiting line, or queue, at a service facility to get a particular service. The speed at which they are served depends on the service rate and the service-time distribution. How long the customers have to wait in line to get the service depends on many things. One of the significant factors is the number of service facilities, or channels, and their sequence.

There are four basic types of queuing situations which affect the waiting time of customers. The simplest situation is known as a single-channel, single-phase case. In this case the customers arrive and form a single line to await service at one service facility. No sequence of service facilities is involved. Another situation is known as the multiple-channel, single-phase case. In this case, the number of service facilities is more than one, and it is possible for arriving customers to wait in several lines to get service. No sequence of service facilities is involved. The third basic type of queuing situation is known as the single-channel, multiple-phase case. In this case, the customers cannot choose from among several service facilities; they must wait for their service from one channel, or service facility. However, once a customer is serviced, he moves on to sequential service facilities. The final basic type of queuing situation is the multiple-channel, multiple-phase case. In this case the customers may form waiting lines at several alternative facilities and, after being serviced at the initial stations, pass on to successive stations.

The reason for breaking down the various queuing situations into four basic types is that the analytical approaches available to study them differ for each type. They also differ with the nature of arrival distributions and service-time distributions. In addition, the analytical approaches differ where the waiting line can reach a definite maximum length (finite waiting lines) and where the waiting line theoretically can become extremely long, approaching infinity (infinite waiting lines).

The systems analyst can effectively utilize queuing theory to find answers to such questions as the following:

1. How many customers will be waiting in line?
2. How long will these customers have to wait for service?
3. How many service facilities should be provided to minimize the costs and time involved?
4. How much idle time will the service facilities have?

These questions and others can be answered with queuing theory. The answers to them provide basic information which will affect management decisions concerning arrival schedules, speeds of service facilities, the number of facilities, and related flow rate questions.

Although there are a number of analytical approaches to different types of queuing problems, we shall not examine all of them here. For illustrative purposes, we shall examine only the single-channel, single-phase case. We shall also make several basic assumptions without getting involved at this point in the underlying mathematics and statistics.

As an example of an application of queuing theory we can return to the question of the loading and unloading facilities for trucks. We shall assume that we want to know how many trucks will be waiting for loading and unloading, how much time will be wasted by having the trucks wait for loading and unloading, and what the probability is that the loading and unloading dock and workers will be idle.

We shall assume that the following assumptions are valid, without getting involved in detailed discussions of their validity.

1. The arrival distribution is a Poisson distribution.
2. The service-time distribution is exponential.
3. Infinite waiting lines are theoretically possible.
4. The queue discipline is first come, first served.
5. The average service rate is greater than the average arrival rate.
6. We are dealing with a single-channel, single-phase case.

Given these assumptions as valid, we can utilize certain equations to find answers to the questions posed above. We must know the average arrival rate and the average service rate to solve these problems. Let us assume that the trucks arrive at the plant every 40 minutes on the average. The average (mean) arrival rate, symbolized by lambda (λ), will be, therefore, 1.5 arrivals per hour. Let us assume that the loading and unloading dock and workers can unload or load a truck every 30 minutes, on the average. Therefore, the service rate, symbolized by mu (μ), will be 2.0 services per hour.

To find the average (mean) number of trucks in the waiting line, we can use the following equation:

$$L_q = \frac{\lambda^2}{\mu(\mu - \lambda)}$$

$$L_q = \frac{(1.5)^2}{(2.0)(2.0 - 1.5)} = \frac{2.25}{1} = 2.25 \text{ trucks waiting in line}$$

To find the average (mean) number of trucks in the waiting line, including the one being serviced, we can use this equation:

$$L = \frac{\lambda}{(\mu - \lambda)}$$

$$L = \frac{1.5}{2.0 - 1.5} = \frac{1.5}{0.5} = 3 \text{ trucks in line, including the truck being serviced}$$

To find the average (mean) waiting time of trucks in line, we can use the equation at the top of page 249.

$$W_q = \frac{\lambda}{\mu(\mu-\lambda)}$$

$$W_q = \frac{1.5}{(2.0)(2.0-1.5)} = \frac{1.5}{1} = 1.5 \text{ hours per truck}$$

To find the average (mean) waiting time of trucks in line, including the truck being serviced, the following equation can be used:

$$W = \frac{1}{\mu-\lambda}$$

$$W = \frac{1}{2.0-1.5} = \frac{1}{0.5} = 2 \text{ hours per truck}$$

The probability that the loading and unloading dock and workers will be idle can be determined by using the following equation, where $n = 0$; that is, the number of trucks being serviced is zero:

$$P_n = \left(1-\frac{\lambda}{\mu}\right)\left(\frac{\lambda}{\mu}\right)^n$$

$$P_n = \left(1-\frac{1.5}{2.0}\right)\left(\frac{1.5}{2.0}\right)^0$$

Since a number other than zero raised to the zero power equals one,

$$P_n = (0.25)(1) = 0.25$$

This indicates that the service facility would be idle one-fourth of the time.

Suppose that the systems analyst finds, in comparing the cost of idle trucks and drivers versus the costs of the idle service facility, that there are too many trucks waiting for too long a period of time. He might suggest that loading and unloading of trucks be standardized and that uniform quantities be carried on each truck. If this is possible, then the service times would become constant instead of exponential. If it is assumed that no other changes in our assumptions are made, the following equations will yield 1) the average (mean) number of trucks in the waiting line and 2) the average (mean) waiting time.

$$L_q = \frac{\lambda^2}{2\mu(\mu-\lambda)}$$

$$L_q = \frac{(1.5)^2}{(2\times2.0)(2.0-1.5)} = \frac{2.25}{2.0} = 1.125 \text{ trucks waiting in line}$$

$$W_q = \frac{\lambda}{2\mu(\mu-\lambda)}$$

$$W_q = \frac{1.5}{(2\times2.0)(2.0-1.5)} = \frac{1.5}{2.0} = 0.75 \text{ hour per truck}$$

The incorporation of constant service times reduces substantially the number of trucks waiting from 2.25 trucks to 1.125. It also reduces the waiting time per truck from 1.5 hours to 0.75 hour.

If it is possible to palletize the loads and use lift trucks, further reductions in waiting time would be possible. Let us assume that if this were done the service time could be reduced to a constant 15 minutes per truck, or four services per hour. The average number of trucks waiting in line would then be reduced to 0.1125.

$$L_q = \frac{(1.5)^2}{(2 \times 4.0)(4.0-1.5)} = \frac{2.25}{20} = 0.1125 \text{ truck waiting in line}$$

The average (mean) waiting time of trucks in line would be reduced to 0.075 hour per truck, or 4.5 minutes.

$$W_q = \frac{1.5}{(2 \times 4.0)(4.0-1.5)} = \frac{1.5}{20} = 0.075 \text{ hour per truck}$$

These examples of the use of queuing theory to examine alternatives are just a few of the many equations available. The various available equations can be of great use to analysts and managers in considering alternatives involving queues. Each type of equation is based on particular assumptions about the problem. Therefore, the validity of the assumptions as they are related to the problem should be determined before particular equations are employed.

In summary, queuing theory provides an analytical mathematical model with probabilistic underpinnings which is applicable to the analysis of differential flow rates. It can be used to determine where in the system differential flow rates will cause disequilibrium and also predict the expected magnitude of disequilibrium. Alternative decision rules with respect to resource allocation or modification of arrival-time distributions or service-time distributions also can be tested to determine their effects on system equilibrium.

Monte Carlo Simulation

Some queuing problems cannot be solved directly with equations. If the arrival rates and service rates are not consistent with standard statistical distributions, it becomes very difficult to evaluate alternatives with equations alone. An effective approach to such problems is simulation, using the Monte Carlo technique.

To demonstrate, the technique will be applied to a problem associated with logistics.[2] Assume that a manager is faced with the question

---

[2]This example is a modification of one which appears in James M. Moore, *Plant Layout and Design* (New York: The Macmillan Company, 1962), pp. 266-70.

of determining the optimum number of trucks to provide in his delivery fleet. This question is of interest to the systems designer since he must consider storage, maintenance, and loading facilities requirements. If the firm operates a large fleet of trucks, the customers will be served rapidly, and little overtime will be required to make all of the necessary deliveries. However, a large fleet of trucks involves a large investment and usually results in excessive idle time for some trucks.

A small fleet of trucks, on the other hand, requires a smaller investment, lower facilities and maintenance requirements, and fewer idle trucks. The small fleet may not be able to make all the necessary deliveries on time, and therefore, excessive overtime may be required. To find an optimal answer to the question of how many delivery trucks should be used in the fleet, it is necessary to find the number of trucks which minimize the costs involved.

The solution to this logistics problem can be found by using the Monte Carlo technique. First, we must gather several types of information which are necessary to solve the problem. The arrival distribution must be determined; it is based on the number of shipments which arrive at the loading platform ready for delivery. The service-time distribution must also be ascertained. In this problem, the service-time distribution refers to the time it takes to make deliveries, and will be stated as the number of deliveries which can be made per day per truck. The costs associated with owning and operating trucks is also a necessary item of information. This cost includes the costs of purchasing the truck, depreciation, interest, maintenance, driver labor cost, and so forth. To determine the optimum number of trucks to use in the delivery fleet, it is also necessary to have some information regarding the costs associated with not being able to make all necessary deliveries on time.

If we were interested only in minimizing the costs of owning and operating trucks, we would be prompted to purchase only enough trucks to keep shipments from piling up in excessive inventories on our loading dock. However, if we purchase too few trucks, it is possible that our customers will become upset by the delays in deliveries and take their business elsewhere. The cost of this eventuality is difficult to ascertain. Thus, to keep this illustration simple, we shall assume that the company policy is to deliver every shipment on the same day that it arrives on the loading dock. If all of the shipments cannot be delivered in a normal work day, overtime deliveries will be required. For our illustration, the overtime cost per truck per day will be used as the penalty cost for having too few trucks. The optimum number of trucks will not eliminate all overtime deliveries, however, since to do so would require excessive truck owning and operating costs. We shall assume the following data to be available to the problem under consideration:

1. The normal arrival distribution of customers (shipments to be delivered) is 300 shipments per day with a standard deviation of 30 shipments per day. That is, the average (mean) arrival rate λ is 300 shipments per day.

2. The service-time distribution (number of shipments which can be delivered) is 60 deliveries per day per truck with a standard deviation of six deliveries per day per truck. This indicates that the average (mean) service rate μ is 60 deliveries per day per truck.

3. The cost of operating a delivery truck is $25.00 per day, including the labor costs, operating costs, depreciation, maintenance, and so forth.

4. The delay-of-delivery cost will be represented by overtime costs. Overtime is assumed to cost $5.00 per truck-hour of overtime operation.

With this basic information, plus random numbers which are available from a random-number table, we are ready to start the Monte Carlo simulation of what is likely to happen, given different fleet sizes. The simulation information is summarized in Table 10-1.

In Table 10-1, the Monte Carlo simulation is carried out for truck fleets ranging from one to seven trucks. Fleet sizes greater than seven could also be tested. The random numbers for deliveries required and deliveries completed represent five days of operation from Monday through Friday. In a real simulation problem, a much longer period of time would be covered. This is where computers become very useful, since they can be programmed to carry out simulation problems using thousands of random numbers.

With reference to the table, column a represents the number of trucks being tested. Column b represents the five days of the week. Column c is composed of random numbers taken from a random number table, which are used in generating data concerning the number of deliveries required in column d. Column e is composed of random numbers taken from a random number table; these numbers are used in computing the expected number of deliveries which will be completed in column f. Column g represents the number of overtime deliveries, column h the overtime costs, and column i the total cost of overtime for each alternative fleet size.

The deliveries required each day are simulated by multiplying the standard deviation (30) times the random number for that day and adding the product to the average arrival rate (300). This can be

TABLE 10-1
Monte Carlo Simulation

a	b	c	d	e	f	g	h	i
No. of Trucks	Day	Random Normal Number	Deliveries Required	Random Normal Number	Deliveries Completed	Overtime Deliveries	Cost of Overtime	Total Overtime Cost
	M	−0.452	286	0.803	64	222	$139	
	T	0.338	310	0.835	65	245	138	
1	W	−0.409	287	−1.161	53	234	177	
	T	1.276	338	−0.218	58	280	193	
	F	0.896	326	0.013	60	266	177	$824
	M	−0.452	286	0.803	129	157	97	
	T	0.338	310	0.835	130	180	111	
2	W	−0.409	287	−1.161	106	181	137	
	T	1.276	338	−0.218	117	221	151	
	F	0.896	326	0.013	120	206	137	633
	M	−0.452	286	0.803	194	92	57	
	T	0.338	310	0.835	195	115	71	
3	W	−0.409	287	−1.161	159	128	97	
	T	1.276	338	−0.218	176	162	110	
	F	0.896	326	0.013	180	146	97	432
	M	−0.452	286	0.803	259	27	17	
	T	0.338	310	0.835	260	50	31	
4	W	−0.409	287	−1.161	212	75	57	
	T	1.276	338	−0.218	234	104	71	
	F	0.896	326	0.013	240	86	57	233
	M	−0.452	286	0.803	324	0	0	
	T	0.338	310	0.835	325	0	0	
5	W	−0.409	287	−1.161	265	22	17	
	T	1.276	338	−0.218	293	45	31	
	F	0.896	326	0.013	300	26	17	65
	M	−0.452	286	0.803	388	0	0	
	T	0.338	310	0.835	390	0	0	
6	W	−0.409	287	−1.161	318	0	0	
	T	1.276	338	−0.218	352	0	0	
	F	0.896	326	0.013	360	0	0	0
	M	−0.452	286	0.803	453	0	0	
	T	0.338	310	0.835	455	0	0	
7	W	−0.409	287	−1.161	371	0	0	
	T	1.276	338	−0.218	410	0	0	
	F	0.896	326	0.013	420	0	0	0

expressed as: column $d = 300 + 30c$. In the case of the Monday figures for a one-truck operation, the column d figure is found as follows:

$$300 + (-0.452 \times 30) = 286$$

The deliveries completed each day are simulated by multiplying the standard deviation (6) times the random number for that day and adding the product to the average service-time rate (60). Then this figure is multiplied times the number of trucks in the fleet. This can be expressed as: column $f = a(60 + 6e)$. In the case of the Monday figures for a one-truck operation, the column f figure is found as follows:

$$1[60 + (0.803 \times 6)] = 64$$

The number of overtime deliveries is the difference between the number of deliveries required and the number of deliveries completed. In the case of the first entry in the table, the deliveries required (286) minus the deliveries completed (64) yields 222 deliveries which must be made on overtime. This high level of overtime activity is due to the fact that we are considering at this point only one delivery truck, which is clearly insufficient. As the number of trucks increases, the number of overtime deliveries decreases, as can be seen by scanning the remainder of the table.

The cost of overtime in column h is found by taking the overtime cost per truck-hour ($5.00) times eight hours per day times the number of trucks times the number of overtime deliveries and dividing this product by the deliveries completed. This can be expressed as: column $h = (\$40ag)/f$. In the case of the first entry in the table, the computation of the column h value is as follows:

$$\frac{\$40 \times 1 \times 222}{64} = \$139$$

The cost of overtime for the entire week being simulated is the total of the daily overtime costs. This is reported in column i.

The costs of fleet sizes ranging from one truck to seven trucks can now be determined from the table. The regular time is found by multiplying the number of trucks times the daily cost of operations ($25.00) times the number of days simulated (5). The overtime costs are found in column i for each alternative fleet size. The combined total costs are shown in Table 10-2.

From the data in the table above, it appears that the optimum fleet size is five trucks. This alternative involves some overtime but minimizes total costs. As mentioned earlier, most real problems involve more variables and many more random numbers for generation of data. Thus, computer simulation becomes very attractive for such

TABLE 10-2
Fleet Size Cost Summary

Number of Trucks	Regular Time Costs	Overtime Costs	Total Costs
1	$1 \times \$25 \times 5 = \$125$	$824	$949
2	$2 \times 25 \times 5 = 250$	633	883
3	$3 \times 25 \times 5 = 375$	432	807
4	$4 \times 25 \times 5 = 500$	233	733
5	$5 \times 25 \times 5 = 625$	65	690
6	$6 \times 25 \times 5 = 750$	0	750
7	$7 \times 25 \times 5 = 875$	0	875

problems. Using the Monte Carlo technique and a computer, one can evaluate a great number of alternatives in finding the optimal solution to logistics problems which do not lend themselves to the queuing equations mentioned earlier in the discussion.

The Transportation Problem[3]

The transportation problem is one method of linear programming. It is useful in problems dealing with allocations of materials from sending points to receiving points. Thus it has many applications in the logistics subsystem. These problems may occur when one must determine how materials should be routed among departments in a plant, how finished goods should be routed to dispersed warehouses or wholesalers, how stored products at wholesaler warehouses should be routed to retailers, and so forth. In a sense, it is applicable to those logistics problems involving flow patterns which converge or diverge between connective points within the system. The objective of the technique is to minimize shipping costs while meeting the demands of the receiving points within the supply limits of the sending points.

As an example, consider a company which has three factories located in three cities and four warehouses in four different cities to which finished products are shipped. We shall assume that a homogeneous product is produced, which can be shipped from any of the three factories to any of the four warehouses. The object will be to minimize shipping costs. The costs associated with shipment, supply, and demand figures are given in Table 10-3.

---

[3]Adapted from Hopeman, *op. cit.*, pp. 186–216.

TABLE 10-3
Transportation Costs

From \ To	Warehouse A	Warehouse B	Warehouse C	Warehouse D	Supply (in units)
Factory 1	$2	$4	$1	$3	30
Factory 2	$8	$2	$6	$5	30
Factory 3	$6	$1	$4	$2	20
Demand (in units)	20	20	30	10	80 Total

The matrix indicates that it costs $2 per unit to ship from factory 1 to warehouse A, $8 to ship from factory 2 to warehouse A, $6 to ship from factory 3 to warehouse A, and so forth. The table also indicates the supply and demand characteristics of the sending and receiving points. Factory 1 has a productive capacity of 30 units per day (or some other selected time period), factory 2 has a productive capacity of 30 units, and factory 3 can produce 20 units. Total production is therefore 80 units per day. The demand patterns of the warehouses are 20 units for A, 20 units for B, 30 units for C and 10 units for D— a total demand of 80 units per day. In this case, supply and demand are equal which simplifies the problem. At a later point we shall consider the approach to be taken when supply and demand are unequal.

In approaching the solution to this problem, units will be assigned to certain warehouses from certain factories. The materials-handling routes will then be evaluated in a systematic manner. If a better solution is possible, the systematic analysis will indicate this. The better solution is implemented, and the same systematic analysis will occur again. This goes on through several cycles until no new arrangement will yield a lower total shipping cost. This last arrangement is the optimal answer and the object of the analytical process.

The northwest corner method, as applied to the transportation problem, derives its name from the fact that product allocations start in the northwest corner (upper left-hand corner) and move to the right and downward in a stair-step fashion. Such an approach is entirely arbitrary with respect to cost effectiveness, however it is a common one where computer programs are being designed to solve the problem. The allocation pattern follows these steps:

1. Start by allocating units in the upper left-hand corner.
2. End by allocating units in the lower right-hand corner.
3. In making each allocation decision, consider whether the supply or demand is smaller and use this amount.

4. Always proceed to exhaust supply and/or demand before moving downward or to the right in the rows and columns of the matrix.

The application of this set of steps to the matrix above yields the allocation pattern in Table 10-4.

TABLE 10-4
Northwest Corner Allocation

To / From	Warehouse A	Warehouse B	Warehouse C	Warehouse D	Supply
Factory 1	20	10			30
Factory 2		10	20		30
Factory 3			10	10	20
Demand	20	20	30	10	80 Total

The next step involves the analysis of alternative allocation patterns. In this analysis, each open cell (cell without units allocated to it) is evaluated to determine if another alternative would save money.

The approach involves testing each open cell by assuming that one unit will be moved into it and adjusting the balance of other occupied cells to maintain equilibrium with respect to the supply and demand parameters. This approach involves the following steps:

1. The open cell being evaluated should be treated as an increasing cost, since products are being moved into it. This is designated by a plus sign.
2. Paths should be constructed from the open cell to an occupied cell in either a vertical or horizontal direction. Diagonal paths are not permitted.
3. Right-angle turns should be made, only occupied cells being used, until the path leads back to the open cell.
4. A completed path must involve an equal number of increased costs (+) and decreased costs (−).
5. The plus and minus signs must balance each other, so that the supply and demand restraints are not exceeded.

The effect of the application of these steps is to test each open cell. As an unoccupied cell becomes occupied for testing, provision must be made for subtracting units in another cell in the row or column involved to balance supply or demand. As that occupied cell is adjusted, another cell in the row or column affected must be similarly adjusted. In each test the cumulative additional costs and cost savings are recorded.

As an example consider cell 2-A. In this case a plus is added to 2-A, a minus is recorded for 1-A, a plus is recorded for 1-B, and a minus is recorded for 2-B. This completes a path from the cell being tested, 2-A, using it and three other occupied cells. Subtracting a unit from 1-A balances demand for warehouse A but creates an imbalance for factory 1. That imbalance is corrected by adding a unit to cell 1-B, but this move creates imbalance for warehouse B. Deleting a unit from cell 2-B corrects this and also creates balance for factory 2. It is this requirement for maintaining balance among the rim requirements (supply and demand) that generates the rules for offsetting plus and minus signs in the analytical process.

After such a path is constructed, the increased costs and decreased costs are accumulated and compared. In the case of cell 2-A, the evaluation reveals an increase in shipping cost of $8 for 2-A and $4 for 1-B or a total increase of $12. The decreased costs are $2 for 1-A and $2 for 2-B for a total of $4. The comparison reveals that shipping from factory 2 to warehouse A would result in an increase in shipping costs of $8 per unit ($12 − $4). Such a decision would be inappropriate considering the objective of cost minimization.

In like manner, the other open cells of the matrix would be evaluated. These include 3-A, 3-B, 1-C, 1-D, and 2-D. A comparison of increased and decreased costs would yield the following:

Cell	Increased Cost	Decreased Cost	Net Change
3-A	16	8	+8
2-A	12	4	+8
3-B	7	6	+1
2-D	9	8	+1
1-D	9	12	−3
1-C	3	10	−7

This comparison indicates that the most advantageous move will be into cell 1-C, which will result in a net saving of seven dollars per unit moved. The next question is how many units should be moved into cell 1-C. The realistic answer is as many as possible. The most units which can be moved are limited to the smallest number in any one cell with a minus sign in the path which was used for evaluation.

In evaluating cell 1-C, a unit was added to 1-C, subtracted from 2-C, added to 2-B, and subtracted from 1-B. This same procedure is used in making the revised allocation. Since 10 units represent the smallest number of units in the path, 10 units will be used where one unit was used in the analysis. Thus 10 units will be added to cell 1-C, 10 units will be subtracted from the 20 already in 2-C, leaving a difference of 10 units in 2-C; 10 units will be added to the 10 units already in cell 2-B, making a total of 20 units; and, finally, 10 units will be subtracted from the 10 units in cell 1-B, leaving zero units in cell 1-B. Note that the supply and demand restraints have not been disturbed. That is, factory 1 still supplies a total of 30 units, factory 2 supplies 30 units, and factory 3 supplies 20 units. The same equilibrium exists between the original and revised matrices concerning the demand requirements of warehouses A, B, C, and D. The revised matrix appears in Table 10-5.

TABLE 10-5
Revised Allocation

To From	Warehouse A	Warehouse B	Warehouse C	Warehouse D	Supply
Factory 1	20		10		30
Factory 2		20	10		30
Factory 3			10	10	20
Demand	20	20	30	10	80 Total

At this point, it is possible to explain why the smallest number in the path must be used for replacement. In any replacement matrix, one new cell is filled and another becomes empty. That is, where 10 units existed in cell 1-B in the original matrix, now there are no units. If a larger number were used, say 20 units, and the appropriate additions and subtractions were carried out, the result for cell 1-B would be a minus 10 units. In real business situations, there is no such thing as negative, or minus, units; they either exist or they do not exist. It is for this reason that the smallest number is used. Note that the smallest number to which this rule applies is one which is subject to a subtraction. That is, the smallest number in the path, which will go to zero under subtraction, is the one which should be used for replacement. There are cases where more than one cell will go to zero. This is known as degeneracy and is handled as a special case. It is described later in the chapter.

We are now at the point where one allocation has been established, open cells have been evaluated, and a revised allocation has been made. This represents the first iteration. Typical transportation problems go through several such cycles, or iterations, before the optimum solution is reached. The example considered here requires just two iterations. In the second iteration, the revised matrix is evaluated. There are six new cells which must be analyzed: 2-A, 3-A, 1-B, 3-B,

1-D, and 2-D. The evaluation of these cells takes place in the same fashion discussed above. The results of the evaluation process are shown below:

Cell	Increased Cost	Decreased Cost	Net Change
1-*B*	10	3	+7
1-*D*	7	3	+4
2-*A*	9	8	+1
3-*A*	7	6	+1
3-*B*	7	6	+1
2-*D*	9	8	+1

When no savings result from the evaluation of the open cells, the optimum solution has been achieved and the total transportation cost has been minimized within the limits imposed by the costs involved and the supply-demand relationships. In this example, the transportation costs are as follows for the 80 units produced and demanded:

Transportation Costs

Cell	Units	Cost	Total
1-*A*	20	$2	$40
2-*B*	20	2	40
1-*C*	10	1	10
2-*C*	10	6	60
3-*C*	10	4	40
3-*D*	10	2	20
Totals	80		$210

This problem is a fairly limited one. Many industrial transportation problems involve matrices with dozens of cells. The computation problem is not difficult, but the number of iterations may be large. In this case, a computer program might be used to handle the mechanical aspects of the computations. The transportation problem is quite useful in the analysis of logistics problems in that it leads to optimum solutions. Even though it appears to be a trial-and-error approach, it does cover all of the alternatives, and the resulting answer does represent the optimum, or lowest possible costs in the circumstances.

In those cases where the supply and demand are not equal, it is necessary to introduce a slack variable. Such a slack variable is assigned shipping costs of zero and the necessary number of units is recorded to bring the supply-demand relationship to equality. For example, if supply exceeds demand by 30 units, then an imaginary warehouse is added to the matrix with 30 units demanded by it. In fact, these 30 units would not be shipped or, perhaps, even produced. In the other case, where demand exceeds supply, an imaginary factory would be added to the matrix with a dummy supply capacity. The warehouse which was assigned these units, after analysis was completed, would have to acquire them from another supplier or do without them.

As an example of unequal supply and demand consider the following:

Factory 1 supply = 120	Warehouse A demand = 70
Factory 2 supply = 100	Warehouse B demand = 80
Factory 3 supply = 140	Warehouse C demand = 110
	Warehouse D demand = 60
Total supply 360	Total demand 320

In this case, the supply (360) exceeds the demand (320). In order to solve the problem, an extra 40 units must be added as slack demand. To do this, an imaginary warehouse will be established to which the extra 40 units will be assigned. The matrix representing this condition appears in Table 10-6.

TABLE 10-6
Transportation Costs with Slack

To From	Ware- house A	Ware- house B	Ware- house C	Ware- house D	Slack Ware- house	Supply
Factory 1	$3	$2	$2	$7	$0	120
Factory 2	$4	$3	$1	$5	$0	100
Factory 3	$5	$6	$3	$2	$0	140
Demand	70	80	110	60	40	360 Total

Using the same evaluative procedure as discussed earlier, this transportation problem can be solved and will yield the optimal allocation pattern shown in Table 10-7.

TABLE 10-7
Optimal Allocation Pattern with Slack

To From	Ware- house A	Ware- house B	Ware- house C	Ware- house D	Slack Ware- house	Supply
Factory 1	40	80				120
Factory 2			100			100
Factory 3	30		10	60	40	140
Demand	70	80	110	60	40	360 Total

The results of the evaluation of the open cells indicate that no alternative allocation will yield savings in transportation costs. Thus, the allocation pattern in Table 10-7 is the optimum one and will yield the lowest possible costs. Note that the slack variable, the imaginary warehouse, completes the matrix so that the problem could be solved, yet it did not alter the real supply-demand parameters. A total of 360 units will still be available from the factories, and a total of 320 units will be demanded and shipped, with the 40 extra units remaining at factory 3.

Another type of special case concerning the transportation problem is that of degeneracy. The second rule used in evaluating the open cells requires that a path be constructed from the open cell being evaluated to an occupied cell in either a vertical or horizontal direction. Rule 3 requires that right-angle turns be made, only occupied cells being used, until the path leads back to the open cell. The degenerate case occurs when these rules cannot be followed. The matrix shown in Table 10-8 exhibits this characteristic.

TABLE 10-8
The Degenerate Case

To From	Warehouse A	Warehouse B	Warehouse C	Warehouse D	Supply
Factory 1	40		60		100
Factory 2		70		50	120
Demand	40	70	60	50	220 Total

As this example suggests, it is impossible to evaluate cell 2-A. Although a plus might be put in 2-A, a minus in 1-A, and a plus in 1-C, there is no place to put a minus sign according to rule 3. If a plus is put in 2-A and a minus in 2-B, then the same problem occurs. There is no cell which can be assigned the next plus sign. This is true for all open cells in this matrix. None of them can be evaluated as the matrix stands.

A simple test for degeneracy is found by using the following formula:

$$R + C - 1 = \text{Occupied Cells}$$

where R = the number of rows and C = the number of columns.

If $R + C - 1$ equals the number of occupied cells in the matrix, the problem is not degenerate. If $R + C - 1$ is more than the number of occupied cells in the matrix, the problem will be degenerate for some paths. In the example just cited, the result of applying this formula is the number five $(2+4-1)$ which exceeds the number of occupied cells. Thus, the problem is degenerate.

Now that a formula is available which will diagnose degeneracy, the important question becomes how to resolve the degeneracy so that the problem can be solved. The method for doing this involves using a symbol in place of a number in one of the open cells. The symbol which is commonly used is delta, which represents a quantity so small that it will not affect supply or demand parameters. When placed in the matrix, it will remain there until subtracted out in a replacement matrix. If it is not subtracted out when the final solution is reached, that is, if it is part of the solution, it is ignored. In deciding where to insert the delta, only one condition must be met — the replacement matrix cannot involve a minus quantity in the cell containing the delta. This would result in subtracting units from zero, and negative quantities, in the real world, are not possible to ship from a factory to a warehouse. With delta inserted in an acceptable open cell, the problem can be solved in the same manner as previously discussed.

Although the northwest corner allocation provides a starting allocation for the solution of transportation problems, it is totally arbitrary and may result in a large number of iterations before the optimal solution is found. To minimize this problem, several approaches have been developed which tend to minimize the number of iterations. One of these is the Vogel approximation method.

Once the matrix relationships are established, the first step in the Vogel method is to determine the difference between the lowest cost and next lowest cost for each row and column of the matrix. After these differences are recorded, the next step is to select the row or column with the largest difference. Then the procedure is to allocate as many units as possible to the low cost cell in the row or column selected. After the units are allocated, the row or column which is completely exhausted by the allocation is crossed out. The procedure is then repeated, overlooking the crossed-out row or column until all rows and columns have been crossed out.

The fundamental logic underlying the method is that units will be allocated to a low cost cell which, if not used during the present iteration, may not be available during the next iteration. This explains why the largest difference between the lowest cost and next lowest cost is the determinant of the cell to be selected. Since this logic provides an allocation pattern which favors low cost cells, the resulting pattern is often close to the optimal solution. The matrix in Table 10-9 reveals how the Vogel approximation method is applied.

TABLE 10-9
Vogel Approximation Method Matrix

To From	Warehouses A	B	C	D	Supply	D_1 D_2 D_3 D_4 D_5
Factory 1	(20) $25	(20) $36	(40) $14	$53	80	11 11 22 17
Factory 2	$74	(30) $42	$38	(30) $44	60	4 4 4 2 2
Factory 3	$50	$65	$72	(30) $23	30	27
Demand	20	50	40	60	170 Total	
D_1	25	6	24	21		
D_2	49	6	24	9		
D_3		6	24	9		
D_4		6		9		
D_5		0		0		

In this example the first evaluation of differences reveals that the largest difference occurs in row 3 (factory 3) at D_1. The low cost cell is 3-D with a demand of 60 and supply of 30. Thirty units are allocated to 3-D and that row is crossed out.

The differences are recomputed at D_2, and column A (warehouse A) represents the largest difference of $49. The low cost cell is 1-A with demand of 20 and supply of 80. Twenty units are allocated to 1-A, and column A is crossed out.

Computing the differences at D_3, no longer considering column A or row 3, results in the largest difference of $24 associated with column C. The low cost cell is 1-C with a demand of 40 and a remaining supply, after the preceding allocation at 1-A, of 60 units. Thus, 40 units are allocated to 1-C, and column C is crossed out.

The differences at D_4 reveal that row 1 is the next to be considered with a difference of $17. Within row 1 the lowest remaining cost is at 1-B. The demand is 50 units and the remaining supply is 20 units; thus, 20 units are allocated to cell 1-B.

The computation of differences now is limited to row 2 and columns B and D. After evaluating these, the largest difference occurs in row 2 and the low cost cell is 2-B. The remaining demand is 30 and the supply is 60, so 30 units are allocated to cell 2-B. The only remaining open cell, 2-D, receives the remaining allocation of 30 units which exhausts both the remaining supply of row 2 and remaining demand of column D. The allocation pattern which results is shown in Table 10-10.

TABLE 10-10
Vogel Allocation Pattern

To / From	Warehouse A	Warehouse B	Warehouse C	Warehouse D	Supply
Factory 1	20	20	40		80
Factory 2		30		30	60
Factory 3				30	30
Demand	20	50	40	60	170 Total

An evaluation of the open cells in this matrix reveals that the optimum allocation has been achieved. The open-cell cost disadvantages are shown below:

Cell	Increased Cost	Decreased Cost	Net Change
2-A	110	67	+43
3-A	130	90	+40
3-B	109	65	+44
2-C	74	56	+18
3-C	152	79	+73
1-D	95	80	+15

Summary of the Logistics Subsystem

The logistics subsystem tracks rates of flow in the materials flow network. This information, coupled with volume tracking in the inventory stock status subsystem, is essential for effective operation of the inventory planning and control subsystem. The tracking points in the system are discussed in this chapter as well as the problems of balance within convergence-divergence patterns.

Several analytical techniques are explored as they pertain to the logistics subsystem. They provide models for decision systems which affect materials flow management. Queuing theory involves analysis of waiting lines or queues. These lines often develop in the materials flow network at points of convergence and at other points where differential flow rates exist between decoupling points or locations of buffer inventories.

Monte Carlo simulation involves analysis of flow problems which are not characterized by arrival rates and service rates which form standard statistical distributions. In these cases, the unique characteristics of the arrival and service distributions can be simulated and the impact on flow processes evaluated. The results of such evaluation can have a significant effect on system design modifications.

The transportation problem is one method of linear programming which is useful in analyzing problems associated with the allocation of materials from a variety of sending points to a variety of receiving points. Within the parameters of the statement of the problem, it yields optimal solutions.

Our next concern is with the two remaining subsystems in the materials flow network. Both the inventory stock status and purchasing subsystems are discussed in Chapter 11.

Conceptual Problems

1. Considering a firm with which you are familiar, where would you locate logistics sensors to track rate of flow in the inventory system? Be specific and attempt to determine the procedures and hardware which would be required to accomplish such flow tracking.

2. Characteristic of logistics subsystems are patterns of convergence and divergence. How would you determine, in an actual situation, the routes which exist throughout the system within a given time frame? What procedure would you develop to update the routing structure as it changes?

3. How would you go about modifying an existing system to balance the rates of flow in the logistics network?

4. Examine the concept of queuing theory. Working back from the equations, what empirical phenomena do you believe support the mathematical expressions of the relationships involved? What environmental factors affect queues which are not considered in the queuing equations? How would you incorporate these factors in systems analysis?

5. With which classes of logistics problems would you propose the use of Monte Carlo simulation? Which classes of logistics problems are better suited to analytical processes utilizing standard statistical distributions? What criteria would you use in making the decision to select a particular statistical distribution or engage in Monte Carlo simulation?

6. The transportation problem is discussed in this chapter as it pertains to the logistics subsystem. Cite other areas of application of the technique in industry.

The Inventory Stock Status
and Purchasing Subsystems

The Inventory Stock Status Subsystem

11

As indicated in Chapter 7, one of the key variables in inventory management is the quantity of any inventory item on hand at a given point in space and time. It is the function of the inventory stock status subsystem to provide this information to the inventory planning and control subsystem. The latter, in turn, provides for evaluation of this data in conjunction with other factors to provide decisions for other components of the materials flow network.

Associated with the tracking of information on quantities available at given points in space and time is the tracking of rates of flow through the physical system. The tracking of rates of flow is carried out by the logistics subsystem discussed in Chapter 10.

Figure 11-1 highlights the points within the materials flow network where changes in quantity levels are sensed. Beginning with vendors, quantity data are loaded in data banks to show how much of particular inventory items are on order. As stock status changes from on-order to in-transit, this information is updated in the data banks. The first sensing point within the firm is at the point where the receiving function is performed. Following this, sensors pick up quantities of inventory items as they are logged into raw materials inventories. As the materials go through the transformation processes involved in production, the quantities of in-process inventories are tracked. At the end of the production process, aggregate quantity levels are logged in finished goods inventories. To signal when inventory items are moving from the plant, sensors are incorporated in the shipping area and the quantities are tracked as they are in transit to customers.

In brief there are five general areas where inventory items tend to accumulate over time: vendor plants, raw materials inventories, in-process inventories, finished goods inventories, and customer inventories. In addition, there are other points within the system where the inventory items tend to be associated with movement over time and space rather than accumulating in one space over time. These areas involve the in-transit functions from vendors to receiving and the in-transit functions from shipping to customers as well as the internal transit functions associated with materials handling.

The latter areas are primarily tracked by the logistics subsystem, although quantities in a state of movement are fed to the inventory stock status subsystem. The former areas constitute the critical areas for data acquisition and reporting on a rapid turn-around basis.

Tracking of Quantities in the Input Phase

The initial stimulus to the vendor is a purchase order from the purchasing subsystem. This is triggered by the stock level of the item in question reaching the reorder point which in turn actuates the transmission of the economic order quantity from the inventory planning and control subsystem to the purchasing subsystem.

Once the order is placed, the quantity on order of the item is retained by the inventory stock status subsystem as signalled from the purchasing subsystem. Once the item moves from the vendor to the company it is transferred to in-transit status. When it is received, a signal from the receiving department indicates the quantity received less any returns, shortages, or overages from the order quantity. In special cases of partial shipments, the portion of the order received is also tracked to indicate what is backlogged against the system on the input side. Final disposition of the input phase is indicated by a signal from raw materials inventories which logs the quantity of the item.

Such tracking at the input phase provides early indicators of system performance which, if unfavorable, may lead to stockouts and requirements for expediting and follow-up. In the system design

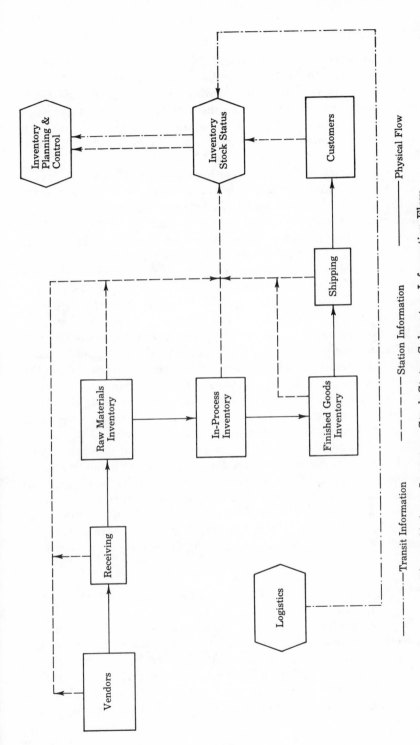

Figure 11-1. Inventory Stock Status Subsystem Information Flow

Transit Information ―――――― Station Information ――――― Physical Flow

phase, the vendor lead times and transit times should be determined; then on the basis of this data, an information loop can be built which will not create managerial reports when the system is functioning normally. Only when the vendor lead time is exceeded and/or transit time extends beyond the standard will exception reports be generated to trigger the expediting and follow-up functions.

Once the items of inventory are logged into raw materials inventory, the replenished stock levels will again exceed the reorder points, and no further signals will be generated by the inventory stock status subsystem to the inventory planning and control subsystem until the inventory levels once again reach reorder points.

If one views the input phase of the inventory stock status subsystem as a closed loop, a circuit can be visualized which involves signals of raw materials stock levels generated daily to be used in comparison tests with reorder points. If the stock level of a particular item is greater than the reorder point, no action is taken. If it is less than or equal to the reorder point, a signal is transmitted to purchasing to place an order for the economic order quantity associated with the item. Once the order is placed, it is tracked from the vendor through receiving to raw materials inventories which, in time, will alter the inventory level at that phase bringing the quantity to a point where daily interrogation will result in no decisions for reorders until the reorder point is reached again.

Such a closed-loop system can be programmed if accurate inputs can be made available from vendors, receiving, and raw materials inventory stations. Although such a program could be designed to operate in real time whenever a change in inventory status occurs in the loop, the expense involved often outweighs the value of the information. For this reason most applications involve batch processing of such data logging and reorder point comparison routines.

Tracking of Quantities in the Process Phase

The process phase between input and output phases can be viewed conceptually as comprising the interfaces of raw materials inventories, in-process inventories, and finished goods inventories as linked by in-plant materials handling functions. The quantity data logging function performed by the inventory stock status subsystem in this phase differs somewhat from its functions at the input and output phases.

As requisitions for raw materials are presented at raw materials inventory stations, corresponding reductions in quantities on hand at those stations can be made. Such withdrawals eventually trigger placement of reorders as discussed above. As with the receiving function, provisions should be made for adjustments other than stock withdrawals. These include shortages encountered when physical inventory counts are taken, deterioration and obsolescence of raw materials inventory items, returns to stock from the production floor of unused raw materials, and the like.

In cases where raw materials require materials handling which consumes a substantial portion of time, move tickets can be used to signal the inventory system that particular quantities of stock items are in transit to work stations. In those cases where employees move the items immediately to a work station, no such tracking occurs since the rate of flow of the material is more rapid than the tracking rate of a reporting system, and the possibility of loss in transit is minimal.

The quantities of items tracked at work stations in the form of in-process inventories present unique problems in the information system. Although a raw material item typically has the same physical characteristics from the vendor to the raw materials inventory and can be easily identified, items in the processing stages change their characteristics as they move through the manufacturing process. Thus a particular raw material item may become a component, later a part to be used in a sub-assembly, then a major part used in final assembly, and finally a finished product.

Raw materials inventories can be broadly classified and their quantities tracked with some ease because of the variety of uses to which they can be put. But as in-process inventories proceed through the production process, the flexibility of their uses in finished products becomes quite restricted. Not only must they be identified as to their physical characteristics but also they require identification with respect to the products with which they are associated. This is particularly evident in job-lot manufacturing. For comparison, consider sheet steel as a raw material. It may be carried in standard grades, gauges, widths, and metallurgical types. By identifying the number of sheets on hand identified by these characteristics, adequate stock level monitoring can be maintained and reorder points can be used effectively to trigger replenishment of stock. If, however, one considers the same steel as in-process inventory, it may take the form of fabricated chair legs which fit only one type of chair and are required in sets of four per finished product. At this point in manufacturing, the descriptions of the inventory items tend to expand substantially, cross-application opportunities diminish, and the critical interface or linkage of information systems tends toward production planning and control with secondary regard for inventory management.

With this in mind, one can see that substantial effort expended in developing flexible coding systems for in-process inventories is well worth while. The payoff comes not only in more effective tracking of in-process inventories but also in terms of more effective production planning and control as well as cost collection.

Tracking of Quantities in the Output Phase

The output phase can be viewed conceptually as being composed of finished goods inventories, the shipping function, and customer inventories. Although most firms concentrate their attention only on the finished goods inventories and shipping function, there is a growing trend toward maintenance of customer stocks on the part of manufacturers. This trend requires an extension of the inventory management function through all channels of distribution to the retail level.

At the finished goods inventory level, two types of conditions typically exist. The first is the maintenance of finished goods inventories which result from producing to stock. The second is the maintenance of finished goods inventories which result from producing to order. The latter condition results in considerably lower average inventory levels than the former. In the case of producing to order, the function of the inventory stock status system is to monitor the level of finished goods and provide data for comparison against order quantities. When the finished goods stock level equals the order quantity required by the customer, a signal is transmitted to shipping which then ships the items to the customer. Quantity tracking in this case signals the start of the shipping function and also monitors the quantity in shipping and in transit to provide information retrieval capabilities when requests for order status are received from customers.

In the case of finished goods resulting from producing to stock, the monitoring of stock levels serves an entirely different function. In such cases, the reorder point concept is linked to the determination of optimum production lot sizes in a manner similar to that associated with reorder points and economic order quantities found at the raw materials inventory stage.

When the quantity of finished goods in inventory equals or is less than the reorder point, a replenishment order is placed with production planning and control which, in turn, results in production activities to produce the optimum production lot size of the item in question. The development of the reorder points, of course, is linked to forecasting models which provide indicators of customer demand over time and approximate usage rates over the short run. Such an information system can be developed as a loop whereby daily interrogation of the inventory stock status subsystem for all stock items yields either no action if the quantity on hand exceeds the reorder point or exception reporting which activates new orders when quantities equal or fall below the reorder point.

The tracking of stock levels in shipping serves two purposes. It indicates the aggregate quantities of units at that station for purposes of customer inquiry and also provides a final check as the finished goods leave the plant that the quantities shipped conform to the quantities ordered by the customers.

In those cases where inventory stock status can be determined at the customer stage, information can be gathered which is very useful for the rest of the system. The initial stimulus to the entire production system comes from the customer. Any incorrect reporting of requirements and significant delays in feedback of information will result in oscillations within the production system which tend to be amplified in magnitude as the effect of the stimulus passes through preceding upstream phases of the system. Therefore, by having quantity sensors at the customer level, both the accuracy of reported usage and the feedback response time can be improved. This, in itself, will substantially dampen such oscillations.

Convergence-Divergence Patterns

In addition to tracking quantities at the critical points in Figure 11-1 of the materials flow network, some consideration must be given to patterns of convergence and divergence within the system. To this point our concern has been with relationships of capacity and volume through tracking via the inventory stock status subsystem and rate of flow tracking through the logistics subsystem. For effective operation, both subsystems must be sensitive to tracking points along convergent and divergent routes through the materials flow network.

Obviously, the convergence-divergence patterns found in business firms take a multitude of forms so that any particular schematic cannot represent a universally applicable pattern. Further, such patterns tend to be exceedingly complex when charted in detail. For these reasons we shall focus attention on a hypothetical construct which is highly simplified yet depicts some fundamental characteristics of many convergence-divergence patterns found in business firms.

Convergence-Divergence in the Input Phase

In Figure 11-2, the schematic depicts the fundamental patterns associated with the input phase of the materials flow network. This pattern exhibits a basic tendency toward convergence although in some cases divergence also exists. It starts with a variety of vendors and ends with the positioning of inventory items in various types of inventories. For simplicity, the inventory items are classified as parts, supplies, and raw materials.

A very common convergence pattern at the input phase involves acquisition patterns from wholesalers who operate regional supply warehouses. In this case the regional supply warehouse would assemble a variety of items from parts, supplies, and raw materials vendors noted as points A, B, and C in Figure 11-2. If such regional supply houses provide effective service to the firm, they can be linked into the materials flow network efficiently. In essence, the firm transfers to them the risks of holding inventories, forecasting demand patterns, and stocking of a variety of items until they are needed. Such an institution is also useful in providing place utility through its regional location. Generally the firm can expect rapid delivery of its orders since it need not go all the way back to the manufacturer which may require extended lead times and transportation times.

Another convergence pattern involves acquisition of parts, supplies, and raw materials from wholesalers who deal only with specific parts, supplies, or raw materials. A primary advantage in this case is the specialized knowledge and specialized product line which the wholesaler can offer to the firm. For example, a wholesaler of steel products usually will have in stock a greater variety of steel products in greater quantities than a general supply house. In addition, since the product line is limited to steel, the firm generally will have sales specialists who can assist in solving technical problems with respect to the choice of the appropriate raw material items. These convergence patterns are noted as points D, E, and F in Figure 11-2.

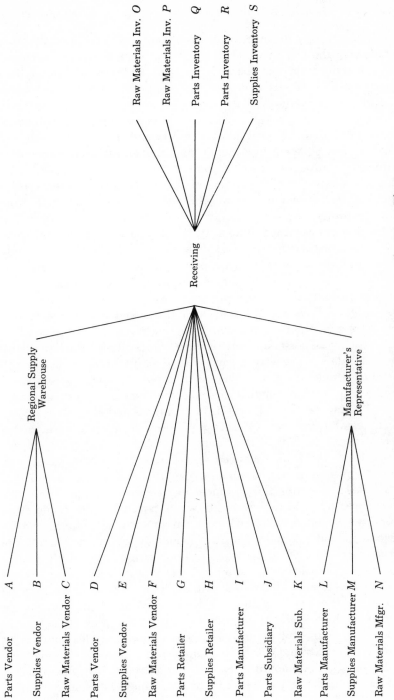

Figure 11-2. Convergence–Divergence Patterns in the Input Phase

Parts Vendor *A*

Supplies Vendor *B*

Raw Materials Vendor *C*

Parts Vendor *D*

Supplies Vendor *E*

Raw Materials Vendor *F*

Parts Retailer *G*

Supplies Retailer *H*

Parts Manufacturer *I*

Parts Subsidiary *J*

Raw Materials Sub. *K*

Parts Manufacturer *L*

Supplies Manufacturer *M*

Raw Materials Mfgr. *N*

Regional Supply Warehouse

Manufacturer's Representative

Receiving

Raw Materials Inv. *O*

Raw Materials Inv. *P*

Parts Inventory *Q*

Parts Inventory *R*

Supplies Inventory *S*

In some cases, the firm acquires inventory items directly from retailers in the city in which it is located. Generally, the inventory items acquired via this channel are parts and supplies rather than raw materials. This is indicated in Figure 11-2 as points G and H. The primary advantage of this type of channel is the rapid delivery which is possible through a local source. Generally the quantities of items purchased are very limited and do not justify the expense of negotiations, competitive bidding, and the formulation of complex purchase contracts. A case in point is the acquisition of office supplies. The primary disadvantage of this pattern is the higher price found at retail in contrast to wholesale or direct purchase from manufacturers.

Another convergence route at the input phase involves acquisition from manufacturers. Figure 11-2 shows this as point I. This route offers unique advantages over others in terms of price and the uniqueness of the product line. Where the product required is to be custom made, it may be the only route available. The disadvantages associated with it are the long lead times and the transportation requirements of moving the item from one manufacturer to another. Since each supplier-manufacturer firm has its own production schedules to balance, the vendee may find that his order is delayed an inordinate amount of time before it is put through the supplier's production process. For this reason, some firms have vertically integrated as indicated in Figure 11-2 by points J and K.

The establishment of subsidiary plants to form an integrated system of materials flow at the input phase offers a number of advantages. First, the firm is assured of a source of the materials, parts, and supplies it needs. Second, it can manipulate the production schedule in the subsidiary to meet its needs rather than the needs of competitors which may occur when purchases are made from other manufacturers. Third, technical process advantages can be developed and maintained to the benefit of the firm. Such trade secrets are often off limits when dealing with other manufacturers. Fourth, a price advantage often accrues to the firm in that several middlemen can be eliminated. Although these advantages may be attractive, there are number of risks in vertical integration as well. For example, a strike at any plant in the system may cause a shutdown of the other plants in the system; such a complex combine of firms can be extremely difficult to manage in an integrated manner; and a substantial amount of capital is required to own manufacturing firms which are part of an input phase materials flow network.

Another common convergence pattern involves manufacturers' representatives. Although these people generally do not provide place utility, they do represent a variety of manufacturers and bring to bear specialized knowledge concerning the solution of specific problems. They are noted in Figure 11-2 at the interface of receiving and points *L, M,* and *N.*

The possibilities for effective systems integration at the input phase vary depending on the convergence route in question. A route from the firm to subsidiary firms offers great potential for such integration. Direct ties to manufacturers provide somewhat less potential, yet long-term contracts can often achieve the kind of integration which can be developed with subsidiaries. Dealing through retailers, wholesalers, and manufacturers' representatives creates some problems. Generally, it is difficult to develop uniform coding systems for inventory items which tie in with the firm's inventory data management system. The development of stock status sensors and logistics sensors with these independent organizations can also be a delicate matter. It is hoped that as these organizations become more aware of the symbiotic nature of their relationship with the firm these obstacles can be overcome. Some progress is already being made in this direction through the adoption of systems contracting, a concept discussed later in this chapter.

Convergence-Divergence in the Process Phase

Once inventory items are received by the firm they are routed over a divergence pattern to a number of stock points such as a number of raw materials inventories, parts inventories, and supply inventories. These are located in various areas of a plant and are noted in Figure 11-2 as points *O, P, Q, R,* and *S.* From those points the items enter the process phase of the materials flow network.

The convergence-divergence patterns in the process phase are depicted in Figure 11-3. The raw materials inventories are noted as points *O* and *P,* the parts inventories as points *Q* and *R,* and the supplies inventory as point *S.* Figure 11-3 is greatly simplified in that convergence-divergence patterns in even very modest manufacturing operations are many times as complex as that indicated in the figure.

The initial divergence patterns begin with raw materials inventories and supply inventories. In Figure 11-3, this is noted as a three-way divergence of raw materials inventory *O* to the three machines in process I, the first phase of manufacturing. Similarly, raw materials inventory P also exhibits a three-way divergence to the same three machines. Since machine operations require certain operating supplies, such as lubricants and tools, the supplies inventory S also diverges to these three machines. The outputs of manufacturing operations in process I are two semi-finished parts, part Y and part Z. In job-lot production, these parts would also constitute inventories serving as buffers between differential production rates.

In the manufacturing activities which constitute process II, machines 4 and 5 draw on raw materials inventories O and P, on the supplies inventory S, and on the semi-finished parts inventories Y and Z. Machine 3, on the other hand, produces a finished part C rather than a semi-finished part in process I. The outputs of process II, which represent further convergence of the material flow network, include part A and part B. When combined with part C from process I and the sub-assembly created from parts inventories Q and R, convergence is completed and consummated in a final assembly process. This process yields the process phase output, the finished product.

Although the schematic in Figure 11-3 indicates, in general terms, the convergence-divergence patterns associated with the process phase, there are a number of other possibilities for convergence and divergence. Parts from a parts inventory may go to one or more machines for further processing. Some semi-finished parts may be rerouted through a given process more than once, creating a cyclical loop. Certain parts may go into the finished product directly without a sub-assembly phase. Often a variety of uniform parts is accumulated at the final assembly phase for assembly into a tremendous variety of finished products rather than the single finished product indicated in Figure 11-3. Such is the case, for example, in the automobile industry where custom and unique finished automobiles exit from the assembly line although the assembly line itself is fed by feeder lines carrying uniform parts.

The tracking of inventory stock status in the process phase is complex, as indicated by the variety of convergence-divergence patterns which exist. Yet, the development of an understanding of the convergence-divergence patterns is essential for effective system design and analysis of systems operation. Fortunately, the interfaces involved in the process phase are all internal to the firm and therefore stock status sensors and logistics sensors can be located at the appropriate inventory points within the firm. They also can be linked through data transmission devices to the firm's computer for dynamic tracking.

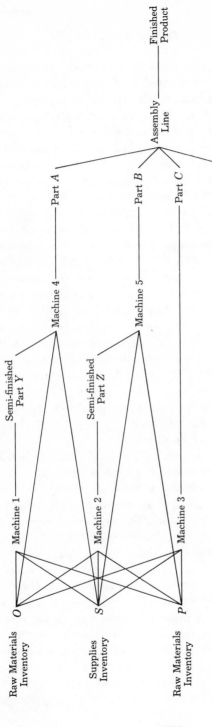

Process I   Process II   Assembly

Machine 1

Machine 2

Machine 3

Machine 4

Machine 5

Semi-finished Part Y

Semi-finished Part Z

Part A

Part B

Part C

Assembly Line

Finished Product

Sub-assembly

Raw Materials Inventory   O

Supplies Inventory   S

Raw Materials Inventory   P

Parts Inventory   Q

Parts Inventory   R

Figure 11-3.   Convergence–Divergence Patterns in the Process Phase

Convergence-Divergence in the Output Phase

The convergence-divergence patterns in the output phase of the materials flow network are indicated in Figure 11-4. In this figure several patterns are indicated which suggest a tendency toward divergence rather than convergence in the output phase. Many institutions in traditional channels of distribution could be cited, but for purposes of simplicity, only a few are indicated in Figure 11-4.

The most common divergence route or channel of distribution is from the firm to wholesalers who, in turn, supply retailers who, in turn, supply customers. This is noted in Figure 11-4 as the route from shipping to wholesaler A who supplies retailers A and B. These retailers supply customers A, B, C and D, E, F respectively.

A somewhat more direct route is indicated in the schematic by the flow from the firm to wholesaler A who in turn, supplies customer G. This is a useful channel if customer G happens to be another manufacturer, an institution — such as an educational institution or hospital — or an individual who has contacts with the wholesaler which allow him to bypass the normal retail outlet.

The most direct route is from the firm directly to the customer. This is indicated in Figure 11-4 as the relationship between the firm and customer H. In this case, customer H may be another manufacturer or an individual who requires a custom-made product which he can negotiate only with the manufacturing firm. It is also found in those cases where the firm sells "seconds," or products with slight defects, through its own factory outlet.

If a firm is located at a significant distance from its markets it may choose to establish a series of regional warehouses to provide greater place utility to its customers. Such an arrangement is indicated in Figure 11-4. One channel involves divergence through a regional warehouse which serves wholesaler B who, in turn, serves retailers C and D who, in turn, serve customers I, J and K, L respectively. Another channel involves the regional warehouse and a direct tie to retailer E who, in turn, serves customers M and N. With this arrangement, the firm carries out the function of wholesaler B with respect to its interface with retailers. The final route in Figure 11-4 involves the firm, its regional warehouse, and customer O, who may be another manufacturer.

The possibility of horizontal and vertical integration exists at the output phase of the materials flow network. The advantages and disadvantages of vertical integration discussed earlier in the chapter apply in this phase as well. In addition, the firm gains significantly

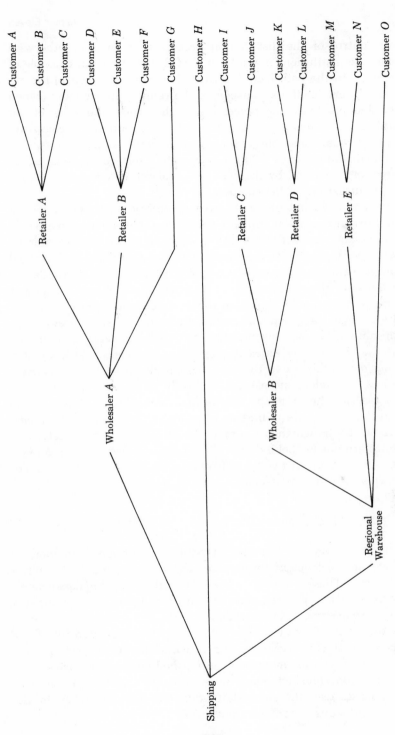

Figure 11-4. Convergence-Divergence Patterns in the Output Phase

more control of the distribution process if it owns or otherwise directs the institutions in its channel of distribution. An example of this exists in the petroleum industry. The refineries control the extraction of crude oil, its transport to the refinery in company-owned ships and pipelines, the processing of the crude into a variety of petroleum products and petrochemicals, distribution via company-owned tankcars and barges to company bulk supply points, and final distribution to a multitude of retail service stations either owned and operated by the refinery or leased to owners who conform to directives of the company.

The establishment of an effective inventory stock status system would involve, at a theoretical level, the establishment of sensors of ultimate customer activity within the output phase of the materials flow network. Since a tremendous number of customers are involved in large firms, such a condition presents a substantial challenge. Ideally, since the entire system is responsive to changes in customer demand, it should be linked to the customers in such a way that these changes can be monitored in real time or, at the very least, with minimal delay. This challenge to existing practice is often dismissed as technically unfeasible by many managers. Yet if we stop to consider it, this is exactly what exists in the provision of electrical services and telephone service. Literally millions of customers in each case are tied into the system and to register a change in demand they need only to flick an electrical switch or lift a telephone receiver. Their demands are met and monitored immediately. The airlines are not far behind with their computerized reservation system — it takes about two seconds to monitor and respond to changes in levels of customer demand.

Summary of the Inventory Stock Status Subsystem

The inventory stock status subsystem is designed to monitor changes in stock levels through sensing at key points in the physical system. These points include vendors; the receiving department; raw materials, in-process, and finished goods inventory stations; the shipping department; and customers.

If the system is viewed conceptually it can be divided into three phases: input (from vendors to raw materials inventories), process (from raw materials inventories to finished goods inventories), and output (from finished goods inventories to customers). Each of these phases uses the stock status quantity monitoring capabilities in different ways for different purposes.

Fundamentally, the inventory stock status subsystem provides two useful outputs when linked to other subsystems. First, it provides raw data for decision models which serve to trigger response patterns within the system. These decision models are used in the inventory planning and control subsystem and in the production planning and control subsystem. Second, the stock status subsystem provides information retrieval capabilities for managers and customers as to the quantity of stock available at any given point in time and space throughout the system.

The logistics subsystem is designed to track rates of flow, an essential item for effective management of the materials flow network. The inventory stock status subsystem is designed to track the volume in the system at critical points. Both of these information inputs are brought together and compared to capacity for studies of modifications in systems design within the inventory planning and control subsystem. They are essential also for purposes of systems analysis when the system is operating over time. In addition to considerations of capacity, volume, and rate of flow, both the logistics subsystem and the stock status subsystem must have sensors established at critical points of convergence and divergence throughout the materials flow network. For purposes of simplicity in developing a convergence-divergence construct to locate these critical points, three linked convergence-divergence networks are developed in this chapter covering the input, process, and output phases of the materials flow network.

The Purchasing Subsystem

The purchasing subsystem is linked with the inventory planning and control subsystem, the inventory stock status subsystem, and the logistics subsystem. It is depicted in Figure 11-5. Its role is to effect the contractual interface between the vendors and the firm. Within the inventory and materials management subsystem it follows particular courses of action relative to the type of item being sought from vendors. In the case of high-cost, high-usage items, the courses of action taken may be rather complex. In the case of low-cost, low-usage items, the purchasing function will be very limited.

Let us consider at the outset the traditional purchasing procedure. The first step is the receipt of purchase requisitions. These are requests made by various personnel within the firm which state what is wanted, how many units are wanted, when the items should be available, and who is making the request. In some cases, the requisitions contain a column headed "Quantity on Hand." This is used to make sure that the person requesting the items has checked the inventory to see if enough items are in stock.

Once the purchase requisitions are received by procurement, the next step is initiated, the analysis of possible sources of supply. The procurement department maintains files on suppliers and can turn to these to compile a list of available suppliers. Those suppliers who are qualified to fill the order are then contacted. If the order is to be determined by bidding, the company will send a request for a price quotation to each supplier. These requests also solicit information concerning discounts, shipping, and delivery dates.

The third step is the analysis of supplier quotations. The bids are reviewed in terms of prices, discounts, shipping, and delivery dates. In addition, the reliability of the supplier, reciprocity, quality of work, and other factors are considered.

The fourth step is the placing of the purchase order. A purchase order is a binding contract if it is accepted by the supplier. Therefore, the procurement department is careful to see that all of the information on the order is accurate. The purchase order should contain descriptions of the items ordered, unit prices, extensions of these prices, quantities ordered, discounts, terms of payment, shipping instructions, the date of the order, the delivery date, a purchase order number, and the signature of the purchasing agent.

The fifth step involves follow-up on the order. On important orders, procurement may check occasionally to determine whether satisfactory progress is being made in filling the order. If a lengthy shipment is involved, the procurement department may check to see that the order is being transported on schedule.

The sixth step is the receipt of the items. When they are received by the receiving department, the items are checked for many particulars, as discussed above. If everything is satisfactory, the last step can be taken.

The final step involves completion of the records. The items are recorded in inventory, the purchase transaction is recorded as complete, and the payment is sent to the supplier.

In some firms this procedure is followed whether the item being purchased is an expensive machine or common office supplies. Under the systems concept, this approach seems to be quite wasteful. In the first place, different purchasing procedures should be used depending on the value and the usage of the item in question — analytical procedures with the high-value, high-usage items and systems contracting with the low-value, low-usage items. We shall now explore each of these alternatives.

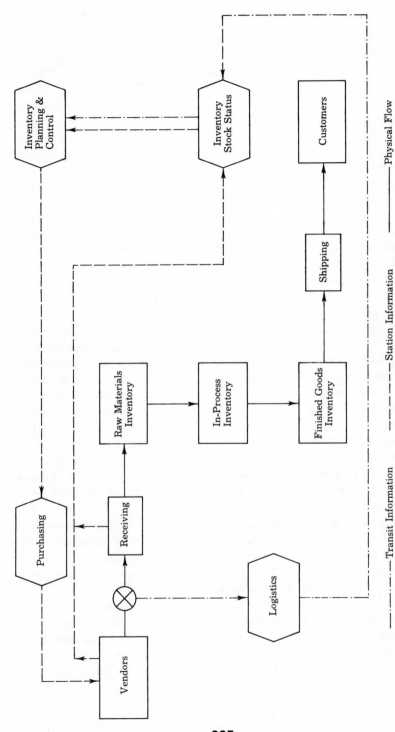

Figure 11-5. Purchasing Subsystem

—————— Physical Flow

— — — — — Station Information

—·—·—·— Transit Information

285

Under the typical purchasing procedure the acquisition of high-value, high-usage items begins with the preparation of purchase requisitions by a wide variety of people throughout the firm. This is generally done when the need arises, which is often too late. Under the systems concept, this kind of information would be automatically triggered from the inventory planning and control subsystem whenever reorder points are reached, regardless of whether anyone happens to notice a shortage or not. The quantity requisitioned, instead of being left to the desires of the requisitioner, is determined in the inventory planning and control subsystem using appropriate economic order quantity models. Instead of putting a fudge factor in the delivery date requested — a common practice under current procedures — the systems approach relies on analysis of reorder point models and the historical and statistical records of past performance on the items in question. These records indicate the reorder point which will cover the expected lead time, and they also have associated with them statistical confidence limits.

The selection of the vendor still remains a major concern under the systems concept; however, the search routines for vendors would very likely be based on information retrieval programs which would pull potential vendors from data banks. Many of the factors involved, such as price, discounts, shipping performance, and other quantifiable factors would be analyzed with models which would yield the list of "best" vendors for a particular item. A subjective overlay would be required, of course, for such factors as reliability of the supplier, reciprocity, and general quality of work.

The generation of purchase orders could be automated once the vendor had been selected and the terms of the contract had been added to those data associated with the requisition forwarded from the inventory planning and control subsystem.

Once the order has been placed and acknowledged by the vendor, the follow-up procedure would be turned over to the logistics subsystem and the stock status subsystem. These two subsystems would track rates of flow from the vendor to the plant as well as conformance to quantity expectations.

The final completion of records would require a signal from receiving to the accounting system for payment. The purchasing subsystem would receive a signal to transfer the purchase order in question from open to closed status, and the inventory stock status subsystem would automatically transfer the item from in-transit status to raw-materials status.

Since the majority of the procedural aspects of traditional purchasing procedures are automated under the systems concept, those people associated with this subsystem can focus more time and attention on the key aspect of their function which requires human skills—selection of the vendor and negotiation with him.

Rather than haggling over terms and price, a group associated with the purchasing subsystem might spend considerably more time engaging in value analysis of the high-cost, high-usage items. Value analysis involves the investigation of an item in terms of its function and price to determine the most effective specifications of the item and to achieve the lowest possible cost. This area of procurement has only recently been exploited. Instead of the purchasing agent's concentrating on the best price for a certain item, his attention is concentrated on the function of the item in value analysis. A value analysis investigation seeks answers to several questions, such as the following:

What is the function of the item and what purpose should it fulfill?
What alternative materials could be used in making the item?
How could the item be simplified?
Could different manufacturing methods be used advantageously?
Could standard, mass-produced parts be used in place of nonstandard parts?
Are there parts in the product which are superfluous to the performance of its function?

Answers to questions such as these often lead to substantial cost savings. The typical procedure used in value analysis is first to examine facts pertaining to costs. These facts are found by examining specifications, drawings, and the products. Next, engineers are consulted to determine if improvements in design can be made. Other people in the firm, as well as customers, may have ideas which will increase the value of the product and lower the cost. In general, at this stage of the procedure, ideas are gathered from all possible sources. Once the ideas are gathered, they are analyzed. When the improved product is envisioned, the vendor is questioned concerning his ability to make the product to include the improvements. At this stage, the engineers of the firm and supplier are brought together to handle the details of the problem. As a final stage of the value analysis procedure, a report is prepared which indicates the savings which will result from the incorporation of the improvements as well as the performance improvements resulting from value analysis.

In order for value analysis to be effective, certain attitudes should prevail among value analysts. The value analyst should think creatively. Even though the traditional approach to making certain products has worked well for years, the value analyst must be able to use his imagination and creativity in finding alternative approaches.

He should be up to date on new products and new processes. Never before has there been such a flood of new products, materials, tools, and supplies on the market. New processes are being developed constantly. Without knowledge of these alternative products and processes, the value analyst is seriously limited in finding the full potential for cost reductions and increases in value received.

When the value analyst reviews the design of a product, it is advisable for him to think in terms of functions. He must determine which functions are necessary and how they might be fulfilled if different materials, tools, supplies, and processes were used. One technique for gaining information in this area involves examining competitive and substitute products to see how they perform the same or similar functions.

As the foregoing implies, the procedural aspects of purchasing are largely automated under the systems concept in purchasing. The area of activity which is retained for those associated with the area is one which is largely analytical.

Now let us turn our attention to the low-volume, low-usage items in inventory. In this area, systems contracting can be applied. Systems contracting involves reasonably long term commitments (two years or more) to a single source for each item. Instead of going through the process of getting competitive bids and selecting vendors, a systems contract is written with one supplier who can supply a class of products, such as general office supplies.[1]

Under this procedure a person who requires replenishment of supplies makes out a three-part requisition which is sent directly to the supplier rather than through purchasing. The supplier then fills the order and delivers it rapidly, retaining one copy of the requisition for his tally of items delivered. When the item is received, one copy serves as the packing list and the other priced copy goes to accounting for accounts payable purposes. At periodic intervals, the vendor reviews his tally of items supplied (every two weeks or every

---

[1]For further information on systems contracting see Ralph A. Bolton, *Systems Contracting—A New Purchasing Technique* (New York: American Management Association, 1966) and *Systems Contracting—A Streamlined Purchasing Technique with Companywide Implications* (New York: American Management Association, 1965).

month) and sends an invoice to the firm. The firm then checks the invoice against the priced requisitions for the period and makes payment to the vendor.

The systems contract concept is one which links the vendor closely to the firm. In many cases this linkage may involve a daily delivery schedule wherein the vendor delivers the items requisitioned the day before and collects the daily requisitions to be filled the next day. In a sense, his relationship with the firm is as close as that relationship within a firm between requisitioners and the firm's storeroom.

This approach has many advantages for particular types of items. It reduces the amount of inventory carried by the firm. This buffer inventory is, in essence, shifted back to the supplier. It markedly reduces the amount of paperwork normally associated with purchasing. It provides additional space which may have been used in the past for storing items which, under the systems contract, are now stored by the vendor. It increases the cash flow by minimizing funds tied up in inventories.

Such a concept is not only beneficial to the firm but also aids the vendor. He is assured of long-range commitments from the firm and can reduce his sales efforts, particularly weekly calls of salesmen. Since he knows that a continuing relationship will exist with the firm, he is better able to manage his own inventories and plan to meet the requirements of the firm. Generally, his paperwork is reduced as is that of the firm; and finally, his cash flow is improved since payment is rapid from the firm when the invoice is presented.

The basic idea, of course, can be automated to some extent. It is conceivable in the future that particular vendors will be linked with their customer firms by data transmission units so that requisitions reach them in minutes. By integrating the data transmission units of the firm and the vendor and tying them to the computer of the firm, it also is conceivable that the inventory stock status and logistics functions could be built into the information stream to provide a dynamic perpetual inventory system and flow rate monitor which remains up to date from hour to hour.

Summary of the Purchasing Subsystem

The discussion of the purchasing subsystem begins with an examination of commonly encountered purchasing procedures as they are used in most firms. The systems approach challenges many of these procedures and many of the functions currently performed are modified and executed by the logistics and stock status subsystems.

The determination of how much to order and when to order is made in the inventory planning and control subsystem. This frees the purchasing subsystem of several decision processes and allows people in this subsystem to concentrate on analytical processes rather than on procedural routines.

One area of analytical effort concerns the performance of the function of value analysis. Value analysis involves the investigation of an item in terms of its function and price to determine the most effective specifications of the item and to achieve the lowest possible cost.

Another area, which emphasizes the symbiotic relationship of suppliers and the firm, is systems contracting. This function is performed within the purchasing subsystem. It involves integration of the supply function and the consumption function in the input phase of the materials flow network.

At this point in the book we have examined all of the subsystems associated with operations management of the materials flow network. The final question we shall address concerns whether or not the concepts presented in this book can be put into practice. Is the systems approach applicable in real-world situations? The answer to this question and its ramifications for implementation are discussed in Chapter 12.

Conceptual Problems

1. Considering a firm with which you are familiar, where would you locate inventory stock status sensors to track quantity levels? Be specific and attempt to determine the procedures and hardware which would be required to accomplish such tracking within the firm.

2. The problems associated with the development of in-process inventory include the coding structure used for the development of the inventory stock status information system. It is compounded by such factors as transformation of raw materials into semi-finished products, parts, subassemblies, and so forth. In addition, provision must be made for coding scrap and rework items. How would you develop a logical classification scheme for such a coding structure?

3. The convergence patterns associated with the input phase of the materials flow network involve a variety of interfaces of the firm to wholesalers, retailers, manufacturers, and manufacturers' representatives. To achieve rapid feedback with respect to stock status what steps could be taken to negotiate the performance of such stock level monitoring on the part of such independent vendors?

4.   Aside from the challenge of instituting a systems approach in the process phase of the materials flow network, job-lot manufacturers must be aware of the location and status of all orders in process. Using a job-lot manufacturing firm with which you are familiar, design an information system which would provide this status information. Consider both the software and hardware ramifications of such a system.

5.   In order to achieve integrated systemic materials flow at the output phase of the materials flow network, some firms have resorted to vertical integration and the ultimate ownership of retailers. What other alternatives are available to achieve such integration where retailers and wholesalers maintain their independence from financial ties to the firm?

6.   If you were to incorporate systems contracting within the purchasing subsystem, how would you modify the existing purchasing organization and alter the responsibilities of those who, prior to the change, were steeped in traditional purchasing procedures.

7.   The systems contracting concept provides a symbiotic linkage of the firm to its immediate suppliers. Could such a concept be extended back from the immediate suppliers to their suppliers or extended forward from the firm to its customers? If so, how should this be done and what repercussions would this have on purchasing and marketing?

Real-Time Systems

Real-Time Systems

Up to this point in the book, we have explored the systems concept and how it was used to achieve major breakthroughs in science; then, based on historical precedents, we have undertaken to explore operations management within a systems framework. We have investigated the environmental set of the firm and the firm itself in a systemic sense. In addition, we have considered the fundamental concepts of systems design and systems analysis.

Following this, in order to explore in some depth the information systems and decision models which can be linked in a systems approach, we evaluated a systemic approach to the management of the materials flow network as one significant subsystem of the firm. Much of this seems promising for the future of management, yet a significant question remains unanswered. Is this promising approach applicable in real-world situations? The answer to this question is, ironically, yes and no.[1]

---

[1] For insights into the controversy over the applicability of the systems concept in business firms see the following: J. A. Beckett, "Wanted — A Concept for Employing Systems Effectively," *Systems and Procedures Journal*, November-December, 1964; W. M. Brooker, "The Total Systems Myth," *Systems and Procedures Journal*, July-August 1965; J. Deardon, "Can Management Information Be Automated?" *Harvard Business Review*, March-April, 1964, and "Myth of Real-Time Management Information," *Harvard Business Review*, May-June, 1966; D. S. Feigenbaum, "Good Systems Are Made Not Born," *Supervisory Management*, August, 1964; A. Harvey, "Systems Can Too Be Practical," *Business Horizons*, Summer, 1964; A. T. Spaulding, "Is the Total System Concept Practical?" *Systems and Procedures Journal*, January, 1964.

Certain aspects of the systems concept as discussed in this book may never be put into practice, or at least not practiced for many decades to come. For example, it has been pointed out that the periodicity of reporting in most firms is based on days, weeks, months, and years. We have seen that the underlying reason for these selected time periods is that they are based on the rotation of the earth about the sun, the rotation of the moon about the earth, and the rotation of the earth on its own axis. These celestial rotation patterns are hardly relevant to business decision making, but still we continue to use them, even though they present awkward time measures in a systemic sense. A cut-over from this system to a more effective system will probably never take place, yet, in some ways, it may be established in the internal information system of the firm.

Another example of a systemic concept which will be difficult to implement in the future concerns organizations. Functionalism has been shown to be non-systemic. The optimization of functional objectives almost always leads to suboptimization for the firm. Yet schools of business administration are departmentalized by functions, students are trained by functions, and many firms are organized by functions. Such a massive "establishment" takes years, if not decades, to change. Harvey addresses this problem in the following way:

> Ultimately, the system solution draws its strength from the application of sound common sense. There is good reason for this. Even the simplest living, growing organism is made up of many parts that interact with each other in complex ways; this is also true of any growing enterprise, in which finance and production, research and development, and marketing are all intertwined. Management has wrenched this entity apart and divided it into functions, authorities, and responsibilities; when the systems approach puts it all together again, it is only restoring to the business its real and inherent unity.[2]

A final example—although many more could be cited to illustrate the difficulty of implementing a systems approach—concerns the need for technological breakthroughs to establish real-time management information systems for a firm. For example, if the materials flow network is viewed as originating with the firm's vendors and ending with the firm's customers, there must be ways of tracking changes in objects of the environmental set as well as within the firm. Although the basic technology is now at hand to achieve this type

---

[2]A. Harvey, "Systems Can Too Be Practical," in P. Schoderbek, *Management Systems—A Book of Readings* (New York: John Wiley & Sons, Inc., 1967), p. 157.

of complex and massive sensing and data transmission, ways must be found to produce stock status and logistics sensors and associated data transmission systems which are inexpensive to acquire and operate. This may take several years.

Now let us turn our attention to the positive answer to the question raised above. Yes, the systems concept *can* be applied in practice. In fact, it has been used effectively in a number of cases which we take for granted. The best illustrations of materials flow networks can be found in the utilities, aerospace, and airline operations.

Telephone Systems

The systems concept is fundamental in the operation of telephone utilities. The objects of the system are telephones, transmission wires, microwave relays, switching circuits, and a complex array of specialized equipment. People are involved, also, as operators, maintenance crews, installers, and so forth. The growth of this system has been so rapid that it outran the number of people who could serve as operators, so no longer do we have a person sending a message communicating with an operator to complete the call, except in special circumstances. The first step in this direction was the dial telephone and associated switching equipment which eliminated the need for an operator on local calls. The next step was direct-distance dialing which eliminated the need for operators on long-distance calls.

Note the similarity to some of the systems concepts discussed earlier. The entire system is integrated; that is, any caller can reach any receiver. It operates in real-time — as demand signals are sent from the customer they are immediately transformed by the firm to meet his demands. Instead of having a production plant, a number of wholesalers, and a number of retailers linking the firm to the customer, the linkage is direct. Each customer can signal a change in demand for the service merely by picking up a telephone and dialing a number. Obviously to meet the needs of such a diverse set of customers, millions of telephones must be available and tied into the system.

To conceive of a telephone system operating in any other way than as an integrated system is to conceive of a chaotic situation. It is reassuring to find that the more systemic the network, the higher its degree of integration and technological sophistication, and the more automated its routine functions have become, the better it serves its customers and increases profits for the firm. Over the years service has improved, rates have declined, even though many material and labor costs have increased, and the system has been able to service a rapidly increasing market which, using the old concepts of telephone service, would never have been obtainable.

Electrical Systems

Electrical systems, those utilities which provide electricity to us, represent another type of familiar real-time system. They represent a unique type of business firm, similar to telephone systems in that they share a unique characteristic—a homogeneous product or service. Both electrical and telephone systems exemplify the concepts discussed in this book.

Consider the vendor-firm-customer interface. An electrical utility is an entirely integrated system. At the vendor or supply side is a source of power, hydroelectric, steam, or atomic, which is transformed into electricity. This electricity is then transported to the customers directly through a vast network of transmission lines and relay stations. Such transmission networks exemplify the concept of integrated convergence-divergence patterns discussed in Chapter 11. Note that the product, in this case, is not storable and thus creates a unique kind of inventory problem — there can be no inventories to act as buffers or decouplers. The system must operate in real time and be customer responsive.

As customers register their changing demand by the flick of a switch or the press of a button, the product is supplied to them immediately. The convergence of these demand change signals back to the producing facility in turn activates it to supply electricity when and where it is needed. As in the case of telephone systems, the electrical systems require sensors of customer demand—millions of them—which are integrated back to the transformation process. The routine "decisions" required in the operation of the system are generally automated and only exceptional cases are reported for managers. Contrast this approach to the alternative of having every customer operate his own electrical generator and it becomes apparent that a systems approach is the only practical way of providing this necessary service. The same analogy can be drawn with the provision of natural gas service, of course.

Aerospace Systems

As we are moving farther in the exploration of space, the value of a systems approach to solving problems is becoming more important.

The objects in a space probe—launch vehicle, on-board navigational equipment, and experiments hardware—represent just a few components of a larger system composed of launch facilities, tracking networks, control systems, and telemetry systems. In the fulfillment of a space probe mission, a unique problem presents itself. Once the launch vehicle has left the pad there can be no manual control of it (except in manned space flights) and its status when in space is unknown, i.e., we cannot examine it on earth. To overcome this difficulty, it is necessary to provide sensors to record the status of the outputs and transformation processes when the vehicle is in flight. It is also necessary to provide standards of performance in terms of a flight program. The sensed characteristics are then compared to the standards in memory, and if necessary due to deviations, effectors are used to modify the status of the vehicle.

Since the operation of a space probe occurs at such a high rate of speed, it is necessary to carry out calculations and make corrections from the ground as well as on board the space vehicle. Therefore telemetry must be developed which will send data from the vehicle back to a receiving and control station on the ground and rapidly return correction signals to the vehicle. This requires sophisticated hardware which operates in real time. It also requires painstaking effort to establish decision rules before the launch to cover every conceivable problem which may arise. The success of the mission, particularly with respect to the control function, rests on the establishment of this information and decision system.

Admittedly, a space probe is quite unlike a business firm and one might ask why this example is presented as a case in support of the practicability of a systems approach. One reason for this is to indicate that we have the technology to meet complex information system problems which could be applied to complex business systems. Although the information systems used in space exploration would be prohibitively expensive for business firms, the firms can rely, for the present, on telephones, teletypes, data transmission systems interfaced with computers, and the like. In the future, even more elaborate and effective hardware should be available to business for this purpose.

Another reason to cite the aerospace example is to emphasize that, where necessity demands it, we have the capability to systemically develop elaborate information and decision systems to meet problems far more challenging than those facing a typical business firm. In aerospace activities it is common to find that after the information and decision systems are designed, a great deal of effort is made to understand the system through simulations. Prior to each actual flight, the mission has been simulated many times. By going through these simulations, potential problems, representing internal systemic impacts or impacts from the environmental set, can be analyzed and solved. If the system fails under one or another of these impacts, the system is redesigned or one or more of the decision models is modified. Such an approach does not abort missions; it simply shows an artificial abort on the simulator. This same approach could be quite useful in the testing of business information systems and decision models used in making management decisions.

Airline Operations Systems

Airline operations represent activities which are characteristic of most firms, or at least more characteristic than the three examples cited above. When viewing the inventory and materials management problems in this industry, one can see that they are extremely complex. Consider what an airline has for sale. It is an empty seat which is available from a point of departure to a point of arrival on a given date at given times. The customer may give the specifications of his order in these terms and may also specify the class of service desired.

This "product," as an inventory item, is complicated not only in terms of its changing identity during the day but also in terms of its extremely short "shelf-life" and the absence of the option to rework or resell it after the shelf-life is exceeded. In essence, the empty seat has a shelf-life of about twenty minutes while the aircraft is at the gate ready for boarding passengers. Once the aircraft moves from the gate to take off, and certainly after takeoff, the value of the empty seat drops to zero and there is no way to recoup this loss. Thus, given the complex problems of short shelf-life, no resale value after take-off, and changing identities during the day, as well as the environmental problems of scheduling with other airlines and modifications of scheduled performance due to weather and congested air traffic, it can be seen that the airlines have a most complex inventory and materials management problem.

Past practices in the airline business were not able to meet growing demands for service and so gave rise to the application of the systems approach. One of the earliest approaches, and one still used by some non-scheduled airlines, was simply to take off only when sufficient passengers showed up to make the trip profitable. This approach would be chaotic if practiced on a large scale today. Another approach was to schedule airline departures, a practice preferred by customers, but to require reservations. Several years ago the making of these reservations was a time-consuming task in that it might be three days or more before a confirmation was received. Even then the airlines tended to oversell because of expected cancellations. Not knowing how many customers would cancel, the airlines occasionally found themselves with irate customers who had tickets but could not board the aircraft because it was filled. This problem has been overcome in a limited way by the introduction of shuttle service where reservations are not needed and the airline continues to supply aircraft departing on a fixed schedule to meet customer demand. At present, such service is limited to a few heavily populated cities, such as the Boston–New York–Washington shuttle service.

By far the most effective solution to this problem was the development of the SABRE system by American Airlines, followed by similar systems introduced by other airlines. This system provides real-time teleprocessing capabilities for data collection and transformation with respect to airline flight reservations.

In the SABRE system, two large-scale digital computers, located at Briarcliff Manor, New York, are linked via terminal interchanges to over one thousand agent sets at ticket counters and reservation centers throughout the country. Similar information linkages provide coverage for international flights for overseas carriers in a number of foreign countries. With this system it is possible for a customer to inquire of the status of a given flight and make a decision concerning the making of a reservation. For example, a customer in Los Angeles may inquire of the American Airlines agent of the status of a flight from Detroit to New York City on a given date and get the answer within two seconds from Briarcliff Manor, New York. The answer may be that the seat desired is available, and if the customer desires it and makes a decision to reserve it, the agent then sends another signal to the central computer confirming the sale. This updates the file on that flight and logs the seat for the particular customer who demanded it, along with other pertinent customer information. Should the customer decide later that he wishes to cancel his reservation, the agent sends this signal to the central computer and thus relieves the inventory log on that flight of one seat of the class specified. In this sense the system is much like the inventory stock status subsystem described in Chapter 11. It keeps track of the current status of a large number of inventories in real time.

The SABRE system and others like it perform other information functions as well. They provide for reservation interfaces with other airlines, process wait-listed passengers, maintain a log of arrival and departure times, and provide messages to agents concerning special action (exception reporting) such as a change in flight status. In such a case, the agent would receive the names and telephone numbers of customers in order to contact them to relay the up-to-date information.

Such a systemic approach, applied to an extremely complex inventory and materials management problem, has proved to be a great stride forward in the operation of airlines. It is, in effect, a large-scale inventory stock status system, which links the demands of thousands of customers to the supplies available, all subject to the dynamic buffeting from environmental impacts that is characteristic of airline operations. This, however, is just part of the application of systems to airline operations. Another significant system concerns the National Airspace Utilization System.

National Airspace Utilization System[3]

Once a passenger is aboard an aircraft, the problem of getting him to his destination safely is one which is of concern not only to the airline but to the national interest. This national concern for air traffic control has led to the development of the National Airspace Utilization System, an advanced system for controlling air traffic.

The operation of this system is comparable to that of the logistics subsystem presented earlier. It is designed to track aircraft from the departure airport, through the enroute traffic, to the arrival airport. It provides not only tracking of rates of flow but also decisions pertinent to the alteration of routes and changes in speed and altitude to promote safe operations. Some of the highlights of this system can be seen as we track a flight from takeoff to landing.

Prior to takeoff, the pilot files a flight plan which is logged into a central computer. The computer checks for errors and provides, with the assistance of a traffic controller, an approved flight plan. This may involve some re-routing due to potential air traffic conflicts which may occur later in time if the flight were to follow the original plan. In essence, the computer serves as a large data bank and dynamic simulator carrying the flight plans of all aircraft in the air as well as those on the ground for which flight plans have been filed. Interfacing the computer are marshalling controllers, departure planners,

---

[3]*Design for the National Airspace Utilization System,* Federal Aviation Agency, Systems Research & Development Service (Washington, D.C.: Government Printing Office, September, 1962).

and ground controllers who monitor the departure time and give instructions to the pilot for taxi routes and takeoff time. These local controllers monitor the flight from the pre-takeoff activity through the actual takeoff from the airport. At this point, control is passed on to air route traffic control centers.

There are several air route traffic control centers established in the United States which cover certain sectors of the country. The size of sectors is dependent on air traffic density and traffic flow patterns. After takeoff, control is passed to departure radar control which routes the aircraft around other departing and incoming aircraft in the vicinity of the departure airport. Once the plane leaves this vicinity, it is monitored by a radar beacon system and other sensors which feed data into the first enroute air route traffic control center. The enroute controller has before him a large radar display unit which shows each aircraft as a blip. The computer provides information to the controller in terms of range and heading, and a transponder on the aircraft provides information concerning altitude. The controller, by comparing the flight plan and actual performance, can then control the flight.

The enroute controller needs only to *monitor* the flight in this system because the computer also tracks actual progress against the flight plan. It will assess continuously the potential for future air traffic congestion which may occur on the flight, and when this seems probable, the computer notifies the enroute controller. He in turn will establish voice communication with the pilot and give instructions for a change in heading or altitude. As the flight progresses, each sector controller hands off the control function to the next sector controller as successive radar scans pick up the aircraft. All the while, the computer monitors the flight and notes changes made by the enroute controller.

As the aircraft approaches its arrival airport, control is passed to a flow control planner who may request a reduction in speed if there is congested air traffic at the airport. The computer monitors this reduction and, through analysis of all other flights around the airport, provides an estimate of arrival clearance time and scheduled landing time. It may also generate a hold signal and determine sequence of aircraft in the stack over the airport. This information is provided to the marshalling controller who will assign a landing time. The planning controller assigns a holding altitude and the aircraft goes into its holding pattern. By monitoring incoming flights, the computer can provide the sequence controller with signals concerning which aircraft should be cleared to land. The sequence controller issues this instruction to the pilot along with information concerning which runway to use, the weather conditions, and descent instructions.

When the aircraft is twenty to twenty-five miles from the runway, control is passed to the final spacing controller. He instructs the pilot concerning the descent of the aircraft in terms of altitude, heading vectors, final time adjustments for arrival, and final approach speed. Once these conditions have been met, control is passed to local tower control. Using radar first and then visual sighting, the local controller brings the aircraft to a safe landing and the ground controller issues appropriate taxi instructions. The computer tracks all of this and, at the completion of the flight, closes the pilot's flight plan.

Note that in this system, on domestic flights, the aircraft is in constant contact with the ground tracking network. Many controllers, radar networks, microwave units, transponder systems, and data processing networks are involved. They provide at all times information concerning the rate of flow of the flight. They also provide for corrective action where necessary. And, of course, because of the high speed involved, the system operates in real time.

Summary

In this chapter we have examined briefly some real-time systems which are in common use. Their existence and effective operation provide evidence that a systems approach to solving complex problems similar to those of business not only does work, but where the problems become large scale and very complex it may be the only answer for effective management.

In each of the cases presented it can be seen that the central concerns are with flow patterns — telephone messages, electricity, and aircraft. Questions of system capacity, volume of flow, and rate of flow are paramount. In each case, elaborate, real-time information systems have been established to provide a managerial analog of the physical system. The manipulation of much of this data is automatic and only exceptions are reported for decisions required of managers.

In each case, vast networks are established to provide for analysis of the integrated system. Within the network, special attention is given to those points where demand is concentrated because of the convergence and divergence patterns which exist at those points. This is exhibited in electrical systems by the wide divergence of electrical lines in a city from the few feeder lines leading to the city and by a similar pattern with respect to telephone systems.

Can such a concept be applied in business? The answer seems to be yes; indeed, many applications have been made. In most cases, these applications have occurred where there is a natural flow process involved such as in petroleum plants, chemical plants, and paper plants. Here, it seems that the natural flow process spawned the managerial system based on flow pattern analysis. That such an approach can be used effectively in other industries as well is just becoming appreciated by managers.

The introduction of computers into such firms has brought about an increased awareness of the systems approach. Since an effective management information system must transcend the mere duplication of existing paperwork procedures, it quickly becomes obvious that as different inputs are added and fewer but more significant outputs are reported, more effective management can emerge.

Conceptual Problems

1. What guidelines could be used to establish appropriate periods of reporting in feedback systems? If these do not conform with current practice of daily, weekly, monthly, and annual reporting, how would you implement the new reporting schedules and still interface effectively with the reporting periodicity required by objects in the environmental set, particularly government?

2. What types of tracking devices could be used for stock status and logistics sensing at critical points of convergence and divergence in the materials flow network? What would you estimate the cost of an information system linking these devices to be? How could this cost be reduced?

3. What are the unique characteristics of electrical, telephone, aerospace, and airline systems which have contributed to the effective development of integrated materials flow systems?

4. The examples of integrated systems cited in this chapter represent large, complex enterprises. Can the systems concept be applied to small firms as well? If so, what guidelines would assure effective implementation of the concept?

5. Although the National Airspace Utilization System represents a sophisticated materials flow system, air traffic at times becomes quite congested at particular airports, notably JFK in New York and O'Hare in Chicago. What factors tend to create this system disequilibrium? What modifications would you recommend to overcome this disequilibrium?

Bibliography

The following bibliography comprises approximately 800 titles associated with the systems concept, systems analysis, and system design. It has been developed to indicate that a rich source of published material is available to those who would pursue the study of systems.

As with any new concept, certain answers and insights are provided to questions which have gone unanswered, or only partially answered, in the past. As old questions are answered, however, more new questions arise with each breakthrough in understanding the new concept. Such is certainly the case with the systems concept. In moving from systems conceptualization and design to the stage of model construction and simulation, one finds that many new questions arise as the process continues. This bibliography may serve as a useful starting point for those seeking answers to these questions.

"Abrasive Maker's Systems Approach Opens New Markets," *Steel*, December 27, 1965.

Abramson, N., *Information Theory and Coding*. New York: McGraw-Hill Book Company, 1963.

Ackoff, R. L., *Progress in Operations Research*. New York: John Wiley & Sons, Inc., 1961.

————, *Scientific Method*. New York: John Wiley & Sons, Inc., 1962.

————, "Systems, Organizations and Interdisciplinary Research," *General Systems*, Vol. 5, (1960).

————, "The Development of Operations Research as a Science," in *Scientific Decision Making in Business*. New York: Holt, Rinehart & Winston, Inc., 1963.

Ackoff, R. L. and P. Rivett, *A Manager's Guide to Operations Research.* New York: John Wiley & Sons, Inc., 1963.

Adams, R. H. and J. L. Jenkins, "Simulation of Air Operation with the Air Battle Model," *Operations Research,* Vol. 8, (1960).

Adamson, Robert E., *Implementing and Evaluating Information Processing Systems.* System Development Corporation Report, SP-1294, November, 1963.

————, *Advances in EDP and Information Systems.* New York: American Management Association, 1961.

Advances in Management Information Systems Techniques. New York: American Management Association, 1962.

Ahrendt, W. R., *Automatic Control.* New York: McGraw-Hill Book Company, 1951.

Ahrendt, W. R., and C. J. Savant, *Servomechanism Practice.* New York: McGraw Hill Book Company, 1960.

Alberts, W. E., "Report of the Eighth AIIE National Conference on the System Simulation Symposium," *Journal of Industrial Engineering,* November-December, 1957.

Allison, H., "Framework for Marketing Strategy," *California Management Review,* Fall, 1961.

Alpert, S. B. and H. Weitz, *Decision Making, Growth, and Failure.* System Development Corporation Report, SP-196, October, 1960.

Amber, G. H. and P. S. Amber, *Anatomy of Automation.* Englewood Cliffs, N.J.: Prentice-Hall, Inc., 1962.

"AMP—System Solution to Manufacturing Planning," *Manufacturing Engineer,* December, 1965.

"Analog Computer—Powerful Tool for Simulation," *Pulp and Paper,* June, 1965.

Anderson, M. W., "The What and Whereto — Management Information Systems," *Total Systems Letter,* September, 1965.

Anderson, T. A., "Coordinating Strategic and Operational Planning," *Business Horizons,* Summer, 1965.

Andrew, G. "An Analytic System Model for Organization Theory," *Academy of Management Journal,* Vol. 8, No. 3, September, 1965.

"A New Look in Management Reporting," *EDP Analyzer,* June, 1965.

Anshen, M., "The Manager and the Black Box," *Harvard Business Review,* November-December, 1960.

Anshen, M. and G. L. Bach, eds., *Management and Corporations 1985.* New York: McGraw-Hill Book Company, 1960.

Ansoff, H. I., *Corporate Strategy.* New York: McGraw-Hill Book Company, 1965.

————, "The Firm of the Future," *Harvard Business Review,* September-October, 1965.

Anthony, R. N., *Planning and Control Systems,* Cambridge, Mass.: Harvard University Graduate School of Business Administration, Division of Research, 1965.

"Applied Psychology for the Systems Man," *Systems and Procedures Journal*, March, 1961.

Arbib, M. A., *Brains, Machines and Mathematics*. New York: McGraw-Hill Book Company, 1964.

Archer, S. H., "The Structure of Management Decision Theory," *Academy of Management Journal*, December, 1964.

Arnoff, L. E., "Operations Research for the Executive," in *Data Processing Yearbook, 1964*. Detroit, Mich.: American Data Processing, Inc., 1963.

Arnoff, L. E. and M. J. Netzorg, "Operations Research — the Basics," *Management Services*, January-February, 1965.

Arnstein, W. E. and E. A. Mock, "A New Approach to the Breakeven Chart," *Management Services*, March-April, 1964.

Arrow, K. J., "Control in Large Organizations," *Management Science*, April, 1964.

————, "Decision Theory and Operations Research," *Operations Research*, December, 1957.

Ash, R. B., *Information Theory*. New York: Interscience Publishers, 1965.

Ashby, W. R., *An Introduction to Cybernetics*. New York: John Wiley & Sons, Inc., 1956.

————, *Design for a Brain*, New York: John Wiley & Sons, Inc., 1952.

Ashar, K. G., "Probabilistic Model of System Operation with a Varying Degree of Spares and Service Facilities," *Operations Research*, Vol. 8, (1960).

Aspects of the Theory of Artificial Intelligence. International Symposium on Biosimulation, New York: Plenum Publishing Corporation, 1962.

"A Total Management System," *Data Processing for Management*, April, 1963.

Bamford, H. E., "Human Factors in Man-Made Machines," *Journal of Human Factors*, November, 1959.

Barbe, E. C., *Linear Control Systems*. Scranton, Pa.: International Textbook Company, 1963.

Barish, N. N., *Systems Analysis for Effective Administration*. New York: Funk and Wagnalls, 1951.

Barlow, R. E. and L. C. Hunter, "Reliability Analysis of a One-Unit System," *Operations Research*, Vol. 9, (1961).

Barnett, J. I., "How to Install a Management Information and Control System," *Systems and Procedures Journal*, September-October, 1966.

Batchelor, J. H., *Operations Research — a Preliminary Annotated Bibliography*. Cleveland, Ohio: The Press of Case Western Reserve University, 1951.

Bates, W. J., "Business Systems in Transition," *Data Processing Magazine*, March, 1963.

Baumann, A. L., "Single Information Flow Philosophy," *Data Processing Yearbook 1963-64*. Detroit, Mich.: American Data Processing, Inc., 1963.

Baumol, W. J., *Economic Theory and Operations Analysis*. Englewood Cliffs, N.J.: Prentice-Hall, Inc., 1965.

Bazovsky, I., *Reliability: Theory and Practice*. Englewood Cliffs, N.J.: Prentice-Hall, Inc., 1961.

Bean, E. E. and W. A. Steger, *Quality Control and Reliability for a Total Weapon System*. Santa Monica, Calif.: The Rand Corporation RM-3130-PR, August, 1962.

Becker, J. and R. M. Hayes, *Information Storage Retrieval*. New York: John Wiley & Sons, Inc., 1963.

Beckett, J. A., "Wanted — A Concept for Employing Systems Effectively," *Systems and Procedures Journal*, November-December, 1964.

Bedford and Onsi, "Measuring the Value of Information, An Information Theory Approach," *Management Services*, January-February, 1966.

Beer, S., *Cybernetics and Management*. New York: John Wiley & Sons, Inc., 1959.

————, "What Has Cybernetics to Do With Operations Research?" *Operations Research Quarterly*, March, 1959.

Bekker, J. A., "Automation: Its Impact on Management," *Advanced Management*, December, 1959.

Bell, D. A., *Information Theory and Its Engineering Applications*. New York: Pitman Publishing Corp., 1962.

————, *Intelligent Machines—An Introduction to Cybernetics*. New York: Blaisdell Publishing Company, 1962.

Bellman, R. E., *Adaptive Control Processes*. Princeton, N.J.: Princeton University Press, 1962.

————, "Control Theory," *Scientific American*, September, 1964.

————, *Dynamic Programming*. Princeton, N.J.: Princeton University Press, 1957.

Bellman, R. E. and R. Kalaba, *Dynamic Programming and Modern Control Theory*. New York: Academic Press, Inc., 1965.

Bendick, M., *The Functional Time Diagram Method of Command and Control System Analysis*. System Development Corporation Report, SP-1111, August, 1963.

Bennet, E. M., *Human Factors in Technology*. New York: McGraw-Hill Book Company, 1963.

Bennett, P. R. *et al.*, *Simpac Users' Manual*. System Development Corporation Report, TM-602-000-00, August, 1963.

Bentley, W. H., "Management Aspects of Numerical Control," *Automation*, October, 1960.

Berg, C. J., "Analyzing an Overall System," *Systems and Procedures Journal*, November-December, 1963.

Berkeley, E. C., *The Computer Revolution*. Garden City, N.Y.: Doubleday & Company, Inc., 1962.

Bernard, E. E. and M. R. Kare, *Biological Prototypes and Synthetic Systems*. New York: Plenum Publishing Corporation, 1962.

Beuter, R. J., "A Theory for the Maintenance of Control of the Firm," *The Journal of Industrial Engineering*, July-August, 1963.

Biegel, J. E., *Production Control—A Quantitative Approach*. Englewood Cliffs, N.J.: Prentice-Hall, Inc., 1963.

Birmingham, H. P. and F. V. Taylor, *A Human Engineering Approach to the Design of Man-Operating Continuous Control Systems*. Washington, D. C.: U. S. Naval Research Laboratory, Report No. 4333, 1954.

Bittel, L. R., *Management by Exception*. New York: McGraw-Hill Book Company, 1964.

Blackwell, D., *Operations Research for Management*. Baltimore, Md.: The Johns Hopkins Press, 1954.

Blank, V. F., "Management Concept in Electronic Systems," *The Journal of Accountancy*, January, 1961.

Blumberg, D. F., "Information Systems and the Planning Process," *Data Processing*, May, 1965.

————, "New Directions for Computer Technology and Applications —A Long Range Prediction," *Computers and Automation*, January, 1964.

Blumenthal, S. C., "Management in Real Time," *Data Processing Magazine*, August, 1965.

Bock, R. H. and W. K. Holstein, *Production Planning and Control— Text and Readings*. Columbus, Ohio: Charles E. Merrill Publishing Company, 1963.

Bode, H. W., *Network Analysis and Feedback Amplifier Design*. New York: D. Van Nostrand Co., Inc., 1945.

Boguslaw, R., *The New Utopians—A Study of System Design and Social Change*. Englewood Cliffs, N.J.: Prentice-Hall, Inc., 1965.

Bohn, E. V., *The Transform Analysis of Linear Systems*. Reading, Mass.: Addison-Wesley Publishing Co., Inc., 1963.

Boldyreff, A. W., *Systems Engineering*. Santa Monica, Calif.: The Rand Corporation, 1954.

Bolton, R. A., *Systems Contracting—A New Purchasing Technique*. New York: American Management Association, 1966.

Bonini, C. P., *Simulation of Information and Decision Systems in the Firm*. Englewood Cliffs, N.J.: Prentice-Hall, Inc., 1963.

Bonney, J. B., "Perceptive Feedback," *Data Processing Magazine*, August, 1964.

Bornemisga, S. T., *The Unified System Concept of Nature*. New York: Vantage Press, 1955.

Bouchardt, R., "The Catalyst in Total Systems," *Systems and Procedures Journal*, May-June, 1963.

Boulding, K. E., "General System Theory—The Skeleton of Science," *General Systems*, Vol. 1, (1956).

Boulding, K. E. and W. A. Spivey, *Linear Programming and the Theory of the Firm*. New York: The Macmillan Company, 1960.

―――, "Towards a General Theory of Growth," *General Systems*. Vol. 1, (1956).

Bourne, C. P., *Methods of Information Handling*. New York: John Wiley & Sons, Inc., 1963.

Bower, J. B. and J. B. Sefort, "Human Factors in System Design," *Management Services*, November-December, 1965.

Bower, J. L. and P. M. Schultheiss, *Introduction to the Design of Servo-mechanisms*. New York: John Wiley & Sons., Inc., 1958.

Bowman, E. H. and R. B. Fetter, *Analysis for Production Management*. Homewood, Ill.: Richard D. Irwin, Inc., 1957.

Boyd, A. W., "Human Relations in Systems Change," *N.A.A. Bulletin*, July, 1959.

Brabb, G. J., "Education for Systems Analysis," *Systems and Procedures Journal*, March-April, 1966.

Brabb, G. J. and E. B. Hutchins, "Electronic Computers and Management Organization," *California Management Review*, Fall, 1963.

Brady, J. S. and S. Goff, *Symbolic Representation of Complex Systems*. Pasadena, Calif.: California Institute of Technology, Jet Propulsion Laboratory, Technical Report No. 32-390, January, 1963.

Brenner, M. E., "A Cost Model for Determining the Sample Size in the Simulation of Inventory Systems," *The Journal of Industrial Engineering*, March, 1966.

―――, "Relation Between Decision Making Penalty and Simulation Sample Size for Inventory Systems," *Operations Research*, May, 1965.

Brewer, S. H., *Rhocrematics — A Scientific Approach to Materials Flows*. Seattle, Wash.: University of Washington Bureau of Business Research, 1960.

Brewer, S. H. and J. Rosenzweig, "Rhocrematics and Organizational Adjustments," *California Management Review*, Spring, 1961.

Bright, J. R., *Automation and Management*. Cambridge, Mass.: Harvard University Graduate School of Business Administration, 1958.

Brillouin, L., *Science and Information Theory*. New York: Academic Press, Inc., 1962.

Brooker, W. M., "The Total Systems Myth," *Systems and Procedures Journal*, July-August, 1965.

Bross, I. D., *Design for Decision*. New York: The Macmillan Company, 1957.

Brown, B. M., *The Mathematical Theory of Linear Systems*. New York: John Wiley & Sons, Inc., 1961.

Brown, G. S. and D. P. Campbell, *Principles of Servomechanisms*. New York: John Wiley & Sons, Inc., 1948.

Brown, J. A., "What New Systems Must Do," *Administrative Management*, January, 1964.

Brown, R. G., *Smoothing, Forecasting and Prediction of Discrete Time Series*. Englewood Cliffs, N.J.: Prentice-Hall, Inc., 1963.

_____, *Statistical Forecasting for Inventory Control*. New York: McGraw-Hill Book Company, 1959.

Bruner, W. G., "Systems Design—A Broader Role of Industrial Engineering," *Journal of Industrial Engineering*, March-April, 1962.

Bruns, R. A. and R. M. Saunders, *Analysis of Feedback Control Systems*. New York: McGraw-Hill Book Company, 1955.

_____, *Automatic Control*. New York: McGraw-Hill Book Company, 1955.

Burck, G., *The Computer and Its Potential for Management*. New York: Harper & Row, Publishers, 1965.

Burlinganie, J. F., "Information Technology and Decentralization," *Harvard Business Review*, November-December, 1961.

Burns, A. L., "Plan Systems Today for Profits Tomorrow," *Office*, January, 1966.

Bursk, E. C. and J. F. Chapman, eds., *New Decision-Making Tools for Managers; Mathematical Programming as an Aid in the Solving of Business Problems*. Cambridge, Mass.: Harvard University Press, 1963.

Business Systems. Systems and Procedures Association, 1965.

Canning, R. G., *Electronic Data Processing for Business and Industry*. New York: John Wiley & Sons, Inc., 1956.

Caples, W. G., "Automation in Theory and Practice," *Business Topics*, Autumn, 1960.

Carasso, M., "Total Systems," *Systems and Procedures Journal*, November, 1959.

Carlson, B., "Industrial Dynamics," *Management Services*, May-June, 1964.

Cass, R. T., "Preventing Resistance to a Systems Change," *N.A.A. Bulletin*, August, 1960.

Celent, C., "Human Factors — Newest Engineering Discipline," *Electron Industries*, February, 1960.

Chamberlain, C. J., "Coming Era in Engineering Management," *Harvard Business Review*, September-October, 1961.

Chang, S. S. L., *Synthesis of Optimum Control Systems*. New York: McGraw-Hill Book Company, 1961.

Chapanis, A., *The Design and Conduct of Human Engineering Studies.* San Diego, Calif.: San Diego State College, 1956.

————, *On Some Relations Between Human Engineering, Operations Research and Systems Engineering.* Baltimore, Md.: The Johns Hopkins Press, 1960.

————, *Research Techniques in Human Engineering.* Baltimore, Md.: The Johns Hopkins Press, 1959.

Chapanis, A. and E. Droz, *Automata,* London: Batsford Ltd., 1958.

Chapanis, A. and T. S. Krawiek, *Systems and Theories of Psychology.* New York: Holt, Rinehart & Winston, Inc., 1960.

Chapanis, A., W. R. Garner and C. T. Morgan, *Applied Experimental Psychology—Human Factors in Engineering Design.* New York: John Wiley & Sons, Inc., 1949.

Chapin, N., *An Introduction to Automatic Computers,* Princeton, N.J.: D. Van Nostrand Co., Inc., 1957.

Chernoff, H. and L. E. Moses, *Elementary Decision Theory.* New York: John Wiley & Sons, Inc., 1959.

Cherry, C., *On Human Communication,* Cambridge, Mass.: The MIT Press, 1965.

Chestnut, H., *Systems Engineering Tools,* New York: John Wiley & Sons, Inc., 1965.

Chin, R., "The Utility of System Models and Developmental Models for Practitioners," in Bennis, W. G., *et. al. The Planning of Change.* New York: Holt, Rinehart & Winston, Inc., 1961.

Chorafas, D. N., *Control Systems, Functions and Programming Approaches.* New York: Academic Press, Inc., 1965.

————, *Systems and Simulation.* New York: Academic Press, Inc., 1965.

Christie, L. S., S. E. Fliege and J. W. Singleton, *Development of Command and Control Systems.* System Development Corporation Report, SP-182, October, 1960.

Christoph, T. G., "Organization of Systems Work — Review and Preview," *Management Services,* May-June, 1966.

Chuang, Y. H., "A Dynamic Programming Model for Combined Production, Distribution, and Storage," *The Journal of Industrial Engineering,* January, 1966.

Churchman C. W., *Prediction and Optimal Decision.* Englewood Cliffs, N.J.: Prentice-Hall, Inc., 1961.

Churchman, C. W. and P. Ratoosh, *Measurement, Definitions and Theories.* New York: John Wiley & Sons, Inc., 1959.

Churchman, C. W. and M. Verhulst, *Management Sciences, Models and Techniques.* New York: Pergamon Press, Inc., 1960.

Churchman, C. W., R. L. Ackoff and E. L. Arnoff, *Introduction to Operations Research.* New York: John Wiley & Sons, Inc., 1957.

Cisler, W. L., "Management's View of the Systems Function," *Systems and Procedures Journal,* July-August, 1965.

Clark, R. N., *Automatic Control.* New York: John Wiley & Sons, Inc., 1962.

Clee, G. H. and F. A. Lindsay, "New Patterns for Overseas Operations — Systems Management Concept," *Harvard Business Review,* January, 1961.

Cleland, D. I., "Why Project Management?" *Business Horizons,* Winter, 1964.

Cline, R. E., *A Survey and Summary of Mathematical and Simulation Models as Applied to Weapon System Evaluation.* Ann Arbor, Mich.: University of Michigan, ASD Technical Report 61-276, October, 1961.

Clough, D. J., *Concepts in Management Science.* New York: Prentice-Hall, Inc., 1963.

Coales, J. F., *Theory of Continuous Linear Control Systems.* London: Butterworths, 1963.

Coch, L. and J. R. French, "Overcoming Resistance to Change," in Cartwright, D. and A. Zander, eds., *Group Dynamics.* Evanston, Ill.: Row, Peterson and Company, 1960.

Coleman, J. J. and I. J. Abrams, "Mathematical Model for Operational Readiness," *Operations Research,* Vol. 10, (1962).

"Coming Changes in Systems Analysis and Design," *EDP Analyzer,* November, 1965.

"Computer-Assisted Corporate Planning," *EDP Analyzer,* September, 1966.

Concepts Associated with System Effectiveness. Bureau of Naval Weapons, Navweps Report 8461, June, 1963.

Conway, B. J. and D. E. Watts, *Business Experience with Electronic Computers.* New York: Controllers Institute Research Foundation, Inc., 1959.

Conway, R. W., *Some Tactical Problems in Simulation Models.* Santa Monica, Calif.: The Rand Corporation, RM-3244-PR, October, 1962.

Cooper, H. G., "The Corporate Information and Data Automation in Perspective," *Systems and Procedures Journal,* May-June, 1966.

Cooper, W. W., "Some Implications of the Newer Analytic Approaches to Management," *California Management Review,* Fall, 1961.

Cosgriff, R. L., *Automatic Control.* New York: McGraw-Hill Book Company, 1958.

————, *Nonlinear Control Systems.* New York: McGraw-Hill Book Company, 1958.

Coughanowr, D. R., and L. B. Koppel, *Process Systems and Analysis Control.* New York: McGraw-Hill Book Company, 1965.

Cox, D. R., *Renewal Theory.* New York: John Wiley & Sons, Inc., 1962.

Coyl, R. J. and J. K. Stewart, "Design of a Real-Time Programming System," *Computers and Automation*, September, 1963.

Culbertson, J. T., "Automation—Its Evolution and Future Direction," *Computers and Automation*, November and December, 1960.

Culliton, J. W., "Age of Synthesis," *Harvard Business Review*, September-October, 1962.

Curtin, K. M., "A Monte Carlo Approach to Evaluate Multi-Moded System Reliability," *Operations Research*, Vol. 7, (1959).

Cyert, R. M. and H. J. Davidson, *Statistical Sampling for Accounting Information*. Englewood Cliffs, N.J.: Prentice-Hall, Inc., 1962.

Cyert, R. M., E. A. Feigenbaum and J. G. March, "Models in Behavioral Theory of the Firm," *Behavioral Science*, April, 1959.

Cyert, R. M. and J. G. March, *A Behavioral Theory of the Firm*. Englewood Cliffs, N.J.: Prentice-Hall, Inc., 1963.

Dalton, M., *Men Who Manage*. New York: John Wiley & Sons, Inc., 1959.

Daniel, D. R., "Management Information Crisis," *Harvard Business Review*, September-October, 1961.

"Data Transmission and the Real-Time System," *Dun's Review and Modern Industry*, September, 1965.

Dauten, P. M., *Current Issues and Emerging Concepts in Management*. New York: Houghton Mifflin Company, 1962.

Davidson, M., *The Use of Simulation in the Development of Man-Machine Systems*. Systems Development Corporation Report, SP-1137, September, 1963.

Davidson, M. and E. Scott, *Simulation Techniques and Their Application*. Systems Development Corporation Report, SP-1133, July, 1963.

D'Azzo, J. J. and C. H. Houpis, *Feedback Control System Analysis and Synthesis*. New York: McGraw-Hill Book Company, 1960.

Dean, B. V., "Application of Operations Research to Managerial Decision Making," *Administrative Science Quarterly*, December, 1958.

Dean, B. V., M. Sasieni and S. K. Gupta, *Mathematics for Modern Management*. New York: John Wiley & Sons, Inc., 1963.

Deardon, J., "Can Management Information Be Automated?" *Harvard Business Review*, March-April, 1964.

————, "How to Organize Information Systems," *Harvard Business Review*, April, 1965.

Dearen, J., *Cost and Budget Analysis*. Englewood Cliffs, N. J.: Prentice-Hall, Inc., 1962.

Dechert, C. R., "The Development of Cybernetics," *The American Behavioral Scientist*, June, 1965.

Deltoro, V., *Principles of Control Systems Engineering*. New York: McGraw-Hill Book Company, 1962.

DeLuca, R., "Introduction to Systems and Procedures," *Systems and Procedures*, January-February, 1961.

Dennis, J. B., *Mathematical Programming and Electrical Networks.* New York: John Wiley & Sons, Inc., 1959.

Desmonde, W. H., *Computers and Their Uses.* Englewood Cliffs, N.J.: Prentice-Hall, Inc., 1964.

DeSpelder, B., "Designing the Participative Decision-Making System," *Systems and Procedures Journal*, January, 1964.

Deutsch, K. W., "Mechanism, Teleology and Mind," *Philosophy and Phenomenological Research*, December, 1951.

_____, "On Communication Models in the Social Sciences," *Public Opinion Quarterly*, Fall, 1952.

"Developments in Fast Response Systems," *EDP Analyzer*, March, 1965.

DiCicco, J. J., "Dynamic Quality Control," *Industrial Quality Control*, November, 1965.

Dickey, E. R., N. L. Senensieb and H. C. Robertson, "An Integrated Approach to Administrative Systems and Data Processing," *Systems and Procedures Journal*, March-April, 1962.

_____, "Automation — The New Technology," *Harvard Business Review*, September-October, 1964.

_____, *Automation — The Advent of the Automatic Factory.* New York: D. Van Nostrand Co., Inc., 1952.

_____, "Automation — The New Technology," *Harvard Business Review*, November-December, 1953.

_____, *Beyond Automation.* New York: McGraw-Hill Book Company, 1964.

_____, "The Application of Information Technology," *Annals of the American Academy of Political and Social Science*, March, 1962.

_____, "What's Ahead in Information Technology?" *Harvard Business Review*, September-October, 1965.

DiRoccaferrera, G. M. F., *Operations Research Models for Business and Industry.* Cincinnati, Ohio: South-Western Publishing Co., 1964.

Doebelorf, E. O., *Dynamic Analysis and Feedback Control.* New York: McGraw-Hill Book Company, 1962.

Doherty, P. A. and J. G. F. Wollaston, "Effective Management of a Systems Project," *Management Services*, March-April, 1965.

Dolansky, L., R. E. Bach, and S. M. Giveen, *Mathematical Analysis of Complex Systems.* Boston, Mass.: Northeastern University, AFCRL 63-38, September, 1962.

Dommasch, D. O. and C. W. Laudeman, *Principles Underlying Systems Engineering.* New York: Pitman Publishing Corp., 1962.

Dorf, R. C., *Time-Domain Analysis and Design of Control Systems.* Reading, Mass.: Addison-Wesley Publishing Co., Inc., 1965.

Dorfman, R., "Operations Research," *American Economic Review,* September, 1960.

―――――, P. A. Samuelson, and R. M. Solow, *Linear Programming and Economic Analysis.* New York: McGraw-Hill Book Company, 1958.

Dorsey, J. T., "A Communication Model for Administration," *Administrative Science Quarterly,* December, 1957.

Dreyfack, R., "Case for Systems Analysis," *Office Management and American Business,* November, 1960.

Drucker, P. F., *America's Next Twenty Years.* New York: Harper & Row, Publishers, 1955.

―――――, *Landmarks of Tomorrow.* New York: Harper & Row, Publishers, 1959.

―――――, "Long-Range Planning, Challenge to Management Science," *Management Science,* April, 1959.

―――――, "Potentials of Management Science," *Harvard Business Review,* January-February, 1959.

―――――, *The Practice of Management.* New York: Harper & Row, Publishers, 1954.

―――――, "Thinking Ahead — Potentials of Management Science," *Harvard Business Review,* January-February, 1959.

Dunlop, J. T., *Automation and Technological Change.* Englewood Cliffs, N.J.: Prentice-Hall, Inc., 1962.

Eckman, D. P., *Principles of Industrial Process Control.* New York: John Wiley & Sons, Inc., 1945.

―――――, "Systems Research and Design," *Proceedings, First Systems Symposium, Case Institute of Technology.* New York: John Wiley & Sons, Inc., 1961.

Edwards, W., "The Theory of Decision Making," *Psychological Bulletin,* July, 1954.

Elkind, J. I., *Characteristics of Simple Manual Control Systems.* Lexington, Mass.: Massachusetts Institute of Technology, Lincoln Laboratories, Report 111, 1956.

Elliott, C. and R. S. Wasley, *Business Information Processing Systems.* Homewood, Ill.: Richard D. Irwin, Inc, 1965.

Ellis, D. O. and F. J. Ludwig, *Systems Philosophy,* Englewood Cliffs, N.J.: Prentice-Hall, Inc., 1962.

Ely, J. H., *Data Collection for Design and Evaluation of Man-Machine Systems.* New York: ASME-58-A-241, December, 1958.

Enrick, N. L., *Management Operations Research.* New York: Holt, Rinehart & Winston, Inc., 1965.

Ernst, M. L., "Operations Research and the Large Strategic Problems," *Operations Research,* July-August, 1961.

Establishing an Integrated Data-Processing System. New York: American Management Association, Special Report 11, 1956.

Etzioni, A., *Complex Organizations.* New York: Holt, Rinehart & Winston, Inc., 1961.

Evans, M. K. and L. R. Hague, "Master Plan for Information Systems," *Harvard Business Review,* January-February, 1962.

Evans, W. R., *Automatic Control.* New York: McGraw-Hill Book Company, 1954.

————, *Control Systems Dynamics.* New York: McGraw-Hill Book Company, 1954.

Evarts, Harry F., *Introduction to PERT.* Boston, Mass.: Allyn & Bacon, Inc., 1965.

Even, A. D., *Engineering Data Processing System Design.* Princeton, N.J.: D. Van Nostrand Co., Inc., 1960.

Ewell, J. M., "How to Organize For a Total System," *Systems and Procedures Journal,* November, 1961.

————, "The Total Systems Concept and How to Organize for It," *Computers and Automation,* September, 1961.

Fabrycyk, W. J. and P. E. Torgersen, *Operations Economy—Industrial Applications of Operations Research.* Englewood Cliffs, N.J.: Prentice-Hall, Inc., 1966.

Fairthorne, R. A., *Towards Information Retrieval.* London: Butterworths, 1961.

Fazar, W., "Progress Reporting in the Special Projects Office," *Navy Management Review,* April, 1959.

Feigenbaum, D. S., "Good Systems Are Made Not Born," *Supervisory Management,* August, 1964.

Feigenbaum, E. A. and J. Feldman, *Computers and Thought.* New York: McGraw-Hill Book Company, 1963.

Feinstein, A., *Foundations of Information Theory.* New York: McGraw-Hill Book Company, 1958.

Ferguson, R. O. and L. F. Sargent, *Linear Programming.* New York: McGraw-Hill Book Company, 1958.

Ferry, W. H., "Caught on the Horn of Plenty," *Bulletin of the Center for the Study of Democratic Institutions,* January, 1962.

Festinger, L. and D. Katz, eds., *Research Methods in the Behavioral Sciences.* New York: Holt, Rinehart & Winston, Inc., 1953.

Fett, G. H., *Feedback Control Systems.* Englewood Cliffs, N. J.: Prentice-Hall, Inc., 1954.

Fetter, R. B. and W. C. Dalleck, *Decision Models for Inventory Management.* Homewood, Ill.: Richard D. Irwin, Inc., 1961.

Finck, N. E., "Line of Balance Gives the Answer," *Systems and Procedures Journal,* July-August, 1965.

Fine, G. H. and P. V. McIsaac, "Simulation of a Time-Sharing System," *Management Science,* February, 1966.

Fisch, G. G., "Line-Staff Is Obsolete," *Harvard Business Review,* September-October, 1961.

Fishburn, P., *Decision and Value Theory*. New York: John Wiley & Sons, Inc., 1964.

Fitchthorn, W. H., "Simulation—A Tool for Management Education," *Systems and Procedures Journal*, January, 1961.

Flagle, C. D., W. H. Huggins and R. H. Roy, eds., *Operations Research and Systems Engineering*. Baltimore, Md.: The Johns Hopkins Press, 1960.

Flehinger, B. J., "System Reliability as a Function of System Age," *Operations Research*, January, 1960.

Flood, M. M., "Operations Research and Automation Science," *Journal of Industrial Engineering*, July-August, 1958.

————, "System Engineering," *Management Technology*, January, 1960.

Floyd, W. F., *Ergonomics and Industry*. London: Institution of Mechanical Engineering, 1958.

Fogel, L. J., *Biotechnology—Concepts and Applications*. Englewood Cliffs, N.J.: Prentice-Hall, Inc., 1963.

Forrester, J. W., *Industrial Design*. New York: John Wiley & Sons, Inc., 1961.

————, *Industrial Dynamics*. New York: John Wiley & Sons, Inc., 1961.

————, "Industrial Dynamics," *Harvard Business Review*, July-August, 1958.

————, *Management and Management Science*. Cambridge, Mass.: Massachusetts Institute of Technology, School of Industrial Management, Memorandum D-48, June, 1959.

————, "Processes Vis-a-vis Systems: Toward a Model of the Enterprise and Administration," *Academy of Management Journal*, March, 1963.

Fowler, R. B., "Analytical Methods in Decision Making," in *Data Processing Yearbook, 1965*. Detroit, Mich.: American Data Processing, Inc., 1964.

Friedman, W. F., "Systems Approach, New Key to Better Mill Profits," *Textile World*, November, 1960.

Fry, B. L. *Applying Electronic Techniques to Management*. System Development Corporation Report, SP-292, April, 1961.

Gaddis, P. O., "The Project Manager," *Harvard Business Review*, May-June, 1959.

Gagne, R. M., *Psychological Principles in System Development*. New York: Holt, Rinehart & Winston, Inc., 1962.

Gagne, R. M. and E. A. Fleishman, *Psychology and Human Performance*. New York: Holt, Rinehart & Winston, Inc., 1959.

Gallagher, J. D., *Management Information Systems and the Computer*. New York: American Management Association, 1961.

Gardner, F. V., *Profit Management and Control*. New York: McGraw-Hill Book Company, 1955.

Gargiulo, G. R., "Network Techniques — A Means for Evaluating Existing Systems," *Systems and Procedures Journal*, November-December, 1964.

Garrott, P. B., "Integrated Data Processing Brings Automation in Paperwork," *Automation*, December, 1954.

Garvin, W. W., *Introduction to Linear Programming*. New York: McGraw-Hill Book Company, 1960.

Gay, L. G., *Stages in the Design and Development of an Information Processing System*. System Development Corporation Report, SP-1023, November, 1962.

Geisler, M. A., *The Sizes of Simulation Samples Required to Compute Certain Inventory Characteristics with Stated Precision and Confidence*. Santa Monica, Calif.: The Rand Corporation, RM-3242-PR, October, 1962.

Gentle, E. C., *Data Communications in Business*. New York: American Telephone and Telegraph Company, 1965.

George, F. H., *Automation, Cybernetics and Society*. London: L. Hill Books, Ltd., 1960.

————, *The Brain as a Computer*. Reading, Mass.: Addison-Wesley Publishing Co., Inc., 1962.

————, *Cybernetics and Biology*. San Francisco: W. H. Freeman and Co., Publishers, 1965.

Gibbs, C. B., *Devices and Developments in Engineering Psychology*. London: Iron and Steel Institute, 1958.

Gibson, J. E. et al., *Control System Components*. New York: McGraw-Hill Book Company, 1969.

Gille, J. C., *Feedback Control Systems*. New York: McGraw-Hill Book Company, 1959.

Ginzberg, E., *Technology and Social Change*. New York: Columbia University Press, 1946.

Glushkov, V. M., *Introduction to Cybernetics*. New York: Academic Press, Inc., 1965.

Gluss, B., "Optimum Policy for Detecting a Fault in a Complex System," *Operations Research*, July, 1959.

Goldberg, J. H., *Automatic Controls—Principles of Systems Dynamics*. Boston, Mass.: Allyn & Bacon, Inc., 1965.

Goldman, S., *Information Theory*. Englewood Cliffs, N.J.: Prentice-Hall, Inc., 1953.

Goldstein, J. R., *Scientific Aids to Decision Making—A Perspective*. Santa Monica, Calif.: The Rand Corporation, 1957.

Goode, H. H. and R. E. Machol, *System Engineering—An Introduction to the Design of Large-Scale Systems*. New York: McGraw-Hill Book Company, 1957.

Goodman, L. L., *Man and Automation*. London: Penguin Books, 1957.

Gordon, R. M., *The Total Information System and the Levels of Data Processing Today—A Progress Report*. New York: American Management Association, 1960.

Gosling, W., *The Design of Engineering Systems*. New York: John Wiley & Sons, Inc., 1962.

Gott, R. C., "Integrating Product and Process," *Automation*, October, 1959.

Gottfried, I. S., "Presenting Complex Systems to Management," *Administrative Management*, August, 1965.

Grabbe, E. M., ed., *Automation in Business and Industry*. New York: John Wiley & Sons, Inc., 1957.

Gray, P. and R. E. Machol, *Recent Developments in Information and Decision Processes*. New York: The Macmillan Company, 1962.

Green, J. C., "Information Explosion, Real or Imaginary," *Science*, May, 1964.

Greenberger, Martin, ed., *Management and the Computer of the Future*. New York: John Wiley & Sons, Inc., 1962.

Greene, J. H., *Production Control—Systems and Decisions*. Homewood, Ill.: Richard D. Irwin, Inc., 1965.

Greenwood, W. T., *Management and Organizational Behavior Theories—An Interdisciplinary Approach*. Cincinnati, Ohio: South-Western Publishing Co., 1965.

Gregory, R. H. and R. L. Van Horn, *Automatic Data-Processing Systems*. Belmont, Calif.: Wadsworth Publishing Co., Inc., 1964.

Greniewski, H. K., *Cybernetics Without Mathematics*. New York: Pergamon Press, Inc., 1960.

Grillo, E. V., "Why Not Practice Human Relations?" *Systems and Procedures Journal*, November, 1964.

Guilbaud, G. T., *What Is Cybernetics?* New York: Criterion Books, Inc., 1959.

Gutenberg, A. W., "A Perspective on Management Control Theory," *Evolving Concepts in Management*. Proceedings of the 24th Annual Meeting of the Academy of Management, Chicago, Ill.: December, 1964.

Haberstroh, C. J., "Controls as an Organization Process," *Management Science*, January, 1960.

Hagstrom, W. O., "Traditional and Modern Forms of Scientific Teamwork," *Administrative Science Quarterly*, December, 1964.

Hall, A. D., *A Methodology for Systems Engineering*. Princeton, N.J.: D. Van Nostrand Co., Inc., 1962.

Hall, A. D. and R. E. Fagen, "Definition of a System," in *Yearbook of the Society for General Systems Research*. New York, 1956.

Haller, G. L., "Information Revolution," *Science Digest*, May, 1964.

Hammerton, J. C., "Automatic Machine Scheduling," *Computers and Automation,* May, 1961.

Hansen, B. L., *Quality Control—Theory and Applications.* Englewood Cliffs, N.J.: Prentice-Hall, Inc., 1963.

Hardie, A. M., *The Elements of Feedback and Control.* London: Oxford Publications, 1964.

Hare, Van Court, *Systems Analysis: A Diagnostic Approach.* New York: Harcourt, Brace & World, Inc., 1967.

Harling, J., "Simulation Techniques in Operations Research—A Review," *Operations Research,* May-June, 1958.

Harman, H. H., *Simulation: A Survey.* System Development Corporation Report, SP-260, July, 1961.

Harrison, H. L., *Control System Fundamentals.* Scranton, Pa.: International Textbook Co., 1964.

—————, *Introduction to Switching and Automata Theory.* New York: McGraw-Hill Book Company, 1965.

Hart, B. L., *Dynamic System Design.* London: Burke Publishing Company, 1964.

Hartmann, H. C., "Management Control in Real-Time Is the Objective," *Systems,* September, 1965.

Harvey, A., "Systems Can Too Be Practical," *Business Horizons,* Summer, 1964.

Hasleton, W. R., "Systems Approach Smooths Start-Up, Increases Handling Efficiency," *Materials Handling Engineering,* March, 1966.

Hatch, T. F., *Proposed Program in Ergonomics.* New York: American Society of Mechanical Engineers, Paper 54-A-238, December, 1954.

Hattery, L. H. and E. M. McCormick, *Information Retrieval Management.* Detroit, Mich.: American Data Processing, Inc., 1962.

Hayes, J. H., "Systems Analysis Concepts," *Military Review,* April, 1965.

Head, R. V., *Real-Time Business Systems.* New York: Holt, Rinehart & Winston, Inc., 1964.

Hemes, R. E., "Management Decisions with the Help of Computers," *Administrative Management,* April, 1966.

Hertz, D. B., "Mobilizing Management Science Resources," *Management Science,* January, 1965.

Heslen, R., "Choosing an Automatic Program for Numerical Control," *Control Engineering,* April, 1962.

Heyne, J. B., *Some Implications of the Analysis of Complex Systems Using Information Flow Techniques.* System Development Corporation Report, SP-501, August, 1961.

Hickey, A. E., "The Systems Approach—Can Engineers Use the Scientific Method," *IRE Transactions on Engineering Management,* June, 1960.

Higgenson, M. V., *Managing with E.D.P.: A Look at the State of the Art*. New York: American Management Association, 1965.

Hill, W. H. and J. H. Wright, "Concept and Design of Integrated Management Information Systems," *Data Processing Yearbook — 1965*. Detroit, Mich.: American Data Processing, Inc., 1964.

Hitch, C. J., "An Appreciation of Systems Analysis," *Operations Research*, November, 1955.

————, *Decision Making for Defense*. Berkeley, Calif.: University of California Press, 1965.

————, *On the Choice of Objectives in Systems Studies*. Santa Monica, Calif.: The Rand Corporation, P-1955, 1960.

Hoag, M., "What Is a System?" *Operations Research*, June, 1957.

Hockman, J., "Specifications for an Integrated Management Information System," *Systems and Procedures Journal*, January-February, 1963.

Hoffman, W., ed., *Digital Information Processors—Selected Articles on Problems of Information Processing*. New York: Interscience Publishers, 1962.

Holt, H. and R. C. Ferber, "The Psychological Transition from Management Scientist to Manager," *Management Science*, April, 1964.

Holzbock, W. G., *Automatic Control — Principles and Practice*. New York: Reinhold Publishing Corp., 1957.

Hoos, I. R., "When the Computer Takes Over the Office," *Harvard Business Review*, July-August, 1960.

Hopkins, L. T., *Integration—Its Meaning and Application*. New York: Appleton-Century-Crofts, 1937.

Hopkins, R. C., "Possible Applications of Information Theory to Management Control," *IRE Transactions on Engineering Management*, March, 1961.

Horowitz, I., *Synthesis of Feedback Systems*. New York: Academic Press, Inc., 1962.

Horrigan, T. J., *Development of Techniques for Prediction of System Effectiveness*. Utica, N.Y.: Rome Air Development Command Publication TDC-63-407, Griffis Air Force Base, February, 1964.

Hosford, J. E., "Measures of Dependability," *Operations Research*, Vol. 8 (1960).

"How to Organize for a Total System," *Systems and Procedures Magazine*, November, 1961.

"How to Use Simulated Systems," *Systems and Procedures Journal*, November, 1959.

Howard, R. A., "Dynamic Programming," *Management Science*, January, 1966.

Hower, R. M. and C. D. Orth, *Managers and Scientists*. Cambridge, Mass.: Harvard Graduate School of Business Administration, Division of Research, 1963.

Hurni, M. L., "Decision Making in the Age of Automation," *Harvard Business Review*, September-October, 1955.

————, "The Basic Processes of Operation Research and Synthesis," in *Scientific Decision Making in Business*. New York: Holt, Rinehart & Winston, Inc., 1963.

————, "The Needs and Opportunities for Operations Research and Synthesis," in *Scientific Decision Making in Business*. New York: Holt, Rinehart & Winston, Inc., 1963.

Impact of Automation. U. S. Dept. of Labor Bulletin 1287, 1960.

"Industrial Simulation — Look Before You Leap," *Iron Age*, December, 1965.

Information Systems Workshop — The Designer's Responsibility and His Methodology. Washington, D. C.: Sparton Books, 1962.

Jackson, J. T., "Information Systems for Management Planning," *Data Processing*, March, 1962.

Jacobson, H. B. and J. S. Roucek, eds., *Automation and Society*. New York: Philosophical Library, Inc., 1959.

Jacoby, J. E. and S. Harrison, "Efficient Experimentation with Simulation Models," *Omega*, TR-60-2, June, 1960.

Jaedicke, R. K., "Current Analytical Techniques Useful in All Decisions," *Stanford Graduate School of Business Bulletin*, Autumn, 1965.

Jasinski, F. J., "Adapting Organization to New Technology," *Harvard Business Review*, January-February, 1959.

Jeffrey, R. C., *The Logic of Decision*. New York: McGraw-Hill Book Company, 1965.

Jerger, J. J., *Systems Preliminary Design*. Princeton, N.J.: D. Van Nostrand Co., Inc., 1960.

Jodka, J., "PERT—A Control Concept Using Computers," *Computers and Automation*, March, 1962.

Johnson, E. A., "The Long-Range Future of Operations Research," *Operations Research*, January-February, 1960.

Johnson, R. A., *Employees — Automation — Management*. Seattle, Wash.: University of Washington, Bureau of Business Research, 1961.

————, "Rhocrematics—A System for Production and Marketing," *Advanced Management*, February, 1961.

Johnson, R. A., F. E. Kast, and J. E. Rosenzweig, "Systems Theory and Management," *Management Science*, January, 1964.

————, *The Theory and Management of Systems*. New York: McGraw-Hill Book Company, 1967.

Jones, M. H., *Executive Decision Making*. Homewood, Ill.: Richard D. Irwin, Inc., 1957.

Joplin, H. B., "The Accountant's Role in Management Information Systems," *Journal of Accountancy*, March, 1966.

Joyce, C. C., *Models: A Method for System Planning*. The Mitre Corporation, ESD-TDR 62-311, July, 1962.

Kagdis, J. and M. R. Lackner, *The Modeling of Management Control*. System Development Corporation Report, SP-1213, December, 1963.

————, "A Management Control Systems Simulation Model," *Management Technology*, December, 1963.

Kami, M. J., "Electronic Data Processing — Promise and Problems," *California Management Review*, Fall, 1958.

Karush, W., *On Mathematical Modeling and Research in Systems*. Systems Development Corporation Report, SP-1039, November, 1962.

Kast, F. E. and J. E. Rosenzweig, "A Survey of the Intra-Company Impact of Weapon System Management," *IRE Transactions on Engineering Management*, March, 1962.

————, "Minimizing the Planning Gap," *Advanced Management*, October, 1960.

————, "Planning — Framework for an Integrated Decision System," *Washington Business Review*, April, 1960.

————, *Science, Technology and Management*. New York: McGraw-Hill Book Company, 1963.

Kaufman, F., "Data Systems That Cross Company Boundaries," *Harvard Business Review*, January-February, 1966.

Kaufmann, A., *Methods and Models of Operations Research*. Englewood Cliffs, N.J.: Prentice-Hall, Inc., 1963.

Kelley, J. E., "Critical Path Planning and Scheduling—Mathematical Basis," *Operations Research*, May-June, 1961.

Kent, A., *Textbook on Mechanized Information Retrieval*. New York: Interscience Publishers, 1962.

Kibbee, J. M., *Management Control Simulation*. System Development Corporation Report, SP-110, July, 1959.

————, *Management Control Systems*. New York: John Wiley & Sons, Inc., 1960.

Kilmer, W. L., *Some Methods for Increasing the Reliability of Complex Digital Communications Systems*. Bozeman, Mont.: Montana State University, TN-59-355, March, 1959.

Klasson, C. R. and K. W. Olm, "Managerial Implications of Integrated Business Operations," *California Management Review*, Fall, 1965.

Klein, M. L. et. al., *Digital Techniques for Computation and Control*. Pittsburgh, Pa.: Instruments Publishers, 1958.

Kleinschrod, W. A., "Testing Decisions in Advance," *Administrative Management*, December, 1964.

Kocher, M., *Some Problems in Information Science*. New York: Scarecrow Press, Inc., 1965.

Koenig, H. E., *Automatic Control*. New York: McGraw-Hill Book Company, 1961.

Koenig, H. E. and W. A. Blackwell, *Electromechanical System Theory*. New York: McGraw-Hill Book Company, 1961.

Kollios, A. E. and J. Stempdi, *Purchasing and EDP*. New York: American Management Association, 1966.

Kompass, E. J., "Information Systems in Control Engineering," *Control Engineering*, January, 1961.

Konvalinka, J. W. and H. G. Trentin, "Management Information Systems," *Management Services*, September-October, 1965.

Koontz, H., "The Management Theory Jungle," *Journal of the Academy of Management*, December, 1961.

Kornfield, L. L. and J. F. O'Hora, "Designing a Management Information System," *Total Systems Letter*, May, 1965.

Kozmetsky, G. and P. Kircher, *Electronic Computers and Management Control*. New York: McGraw-Hill Book Company, 1965.

Kou, B. C., *Automatic Control Systems*. Englewood Cliffs, N.J.: Prentice-Hall, Inc., 1962.

Kraft, J. A., "Human Engineering — New Industrial Tool," *Manufacturers Record*, February, 1958.

Krasnow, H. S. and R. A. Merikallio, "The Past, Present, and Future of Simulation Languages," *Management Science*, November, 1964.

Kushner, A., "Systems Planning," *Bests Insurance News*, November, 1965.

Labo, G. V. and L. M. Benningfield, *Control Systems Theory*. New York: The Ronald Press Company, 1962.

Lach, E. L., "Total Systems Concept," *Systems and Procedures Journal*, November, 1960.

Lackner, *SIMPAC: A Research Tool for Simulation*. System Development Corporation Report, SP-228, March, 1961.

————, *SIMPAC: Toward a General Simulation Capability*. System Development Corporation Report, SP-367, August, 1961.

————, *The Dependency of a Simulation Language on a Theory of Systems*. System Development Corporation, TM-602-400-00, January, 1963.

Laden, H. N. and T. R. Gildersleeve, *System Design for Computer Applications*. New York: John Wiley & Sons, Inc., 1963.

Lange, O. R., *Wholes and Parts — A General Theory of System Behavior*. New York: Pergamon Press, Inc., 1965.

Langill, A. W., *Automatic Control*. Englewood Cliffs, N.J.: Prentice-Hall, Inc., 1965.

Lanning, J. H. and R. H. Battin, *Random Processes in Automatic Control*. New York: McGraw-Hill Book Company, 1956.

Lathi, B. P., *Signals, Systems and Communications*. New York: John Wiley & Sons, Inc., 1965.

Latil, P., *Thinking By Machine — A Study of Cybernetics*. Boston, Mass.: Houghton-Mifflin Company, 1957.

Lawson, C. L., *The Total Systems Concept — Its Application to Manufacturing Operations*. New York: American Management Association, Report No. 60, 1961.

Lawson, W. H., "Computer Simulation in Inventory Management," *Systems and Procedures Journal*, May-June, 1964.

Lazzaro, V., *Systems and Procedures — A Handbook for Business and Industry*. Englewood Cliffs, N.J.: Prentice-Hall, Inc., 1959.

Leavitt, H. J., "Some Effects of Certain Communication Patterns on Group Performance," *Journal of Abnormal Social Psychology*, Vol. 46, (1951).

Leavitt, H. J. and T. L. Whisler, "Management in the 1980's," *Harvard Business Review*, November-December, 1958.

LeBreton, P. P. and D. A. Henning, *Planning Theory*. Englewood Cliffs, N.J.: Prentice-Hall, Inc., 1961.

Ledgerwood, B. *et. al.*, *Control Engineering*. New York: McGraw-Hill Book Company, 1957.

Ledley, R. S., *Digital Computer and Control Engineering*. New York: McGraw-Hill Book Company, 1960.

Leitmann, G., *Optimization Techniques, with Applications to Aerospace Systems*. New York: Academic Press, Inc., 1962.

Leitner, R. G., *A Pragmatist's Approach to System Engineering*. System Development Corporation Report, SP-175, (March, 1961).

Leslie, J. T., "Management's Captive Consultants," *Systems and Procedures Journal*, January-February, 1963.

Letov, A. M., *Stability in Non-Linear Control Systems*. Princeton, N.J.: Princeton University Press, 1961.

Levenstein and Caruthers, *Adaptive Control Systems Symposium*. New York: Pergamon Press, Inc., 1963.

Levy, F. K., "Adaptation in the Production Process," *Management Science*, April, 1965.

Likert, R., *New Patterns of Management*. New York: McGraw-Hill Book Company, 1961.

Limberg, H., "What Management Requires from the Systems Specialists and from the System Itself," *Systems and Procedures Journal*, January-February, 1962.

Lindorff, D. P., *Theory of Sampled-Data Control Systems*. New York: John Wiley & Sons, Inc., 1965.

Lindsey, F. E., *New Techniques for Management Decision Making.* New York: McGraw-Hill Book Company, 1958.

Lipstreu, O., "Organizational Implications of Automation," *Journal of the Academy of Management,* August, 1960.

Litterer, J. A., "The Stimulation of Organizational Behavior," *Journal of the Academy of Management,* April, 1962.

Livingston, J. S., "Decision Making in Weapons Development," *Harvard Business Review,* January-February, 1958.

————, "Weapon System Contracting," *Harvard Business Review,* July-August, 1959.

Lloyd, D. K. and M. Lipow, *Reliability: Management, Methods, and Mathematics.* Englewood Cliffs, N.J.: Prentice-Hall, Inc., 1961.

Lorens, C. S., *Flowgraphs: For the Modeling and Analysis of Linear Systems.* New York: McGraw-Hill Book Company, 1964.

Luce, R. D. and H. Raiffa, *Games and Decisions.* New York: John Wiley & Sons, Inc., 1957.

Lumsdaine, A. A., *Human Factors Methods for System Design.* Pittsburgh, Pa.: The American Institute of Research, 1960.

Lynch, W. A. and J. G. Truxal, *Introductory System Analysis — Signals and Systems in Electrical Engineering.* New York: McGraw-Hill Book Company, 1961.

McCameron, F. A., "Setting Inventory Reorder Points," *Management Services,* May-June, 1965.

McCloskey, J. E., *Operations Research for Management.* Baltimore, Md.: The Johns Hopkins Press, 1954.

McCollom, I. N. and A. Chapanis, *A Human Engineering Bibliography.* San Diego, Calif.: San Diego State College Foundation, 1956.

McCormick, E. J., *Human Engineering.* New York: McGraw-Hill Book Company, 1957.

McDonough, A. M., *Information Economics and Management Systems.* New York: McGraw-Hill Book Company, 1963.

McDonough, A. M. and L. J. Garrett, *Management Systems.* Homewood, Ill.: Richard D. Irwin, Inc., 1965.

MacFarlane, A. G. and G. G. Harrap, *Engineering Systems Analysis.* Reading, Mass.: Addison-Wesley Publishing Co., Inc., 1965.

McGarr, C. J., "Definitive Operating Concept for Service Organizations," *Systems and Procedures Journal,* May, 1965.

McGarrah, R. E., *Production and Logistics Management.* New York: John Wiley & Sons, Inc., 1963.

McGrath, J. D., P. G. Nordlie and W. S. Vaughn, "A Systematic Framework for Comparison of System Research Methods," *Human Science Research,* November, 1959.

McGregor, D., *The Human Side of Enterprise.* New York: McGraw-Hill Book Company, 1960.

McGuire, W. J., "Operations Research in Management Planning and Control," *Journal of Industrial Engineering*, July-August, 1959.

Machol, R. E., ed., *Information and Decision Processes*. New York: McGraw-Hill Book Company, 1960.

————, *System Engineering Handbook*. New York: McGraw-Hill Book Company, 1965.

Machol, R. E. and P. Gray, *Recent Developments in Information and Decision Processes*. New York: The Macmillan Company, 1962.

McKean, R. N., *Efficiency in Government Through Systems Analysis*. New York: John Wiley & Sons, Inc., 1958.

McMillan, C. and R. F. Gonzalez, *Systems Analysis—A Computer Approach to Decision Models*. Homewood, Ill.: Richard D. Irwin, Inc., 1965.

Macmillan, R. H., *Automation*. London: Cambridge University Press, 1956.

————, *Non-Linear Control Systems Analysis*. Oxford, England: Headington Hall, 1962.

McRainey, J. H. and L. D. Miller, "Numerical Control," *Automation*, August, 1960.

Magee, J. F., "The Logistics of Distribution" *Harvard Business Review*, July-August, 1960.

Mahler, W. R., "Systems Approach to Managing by Objectives," *Systems and Procedures Journal*, September, 1965.

Malcolm, D. G., "System Simulation—A Fundamental Tool for Industrial Engineering," *Journal of Industrial Engineering*, May-June, 1958.

————, "The Use of Simulation in Management Analysis—A Survey and Bibliography," *Operations Research*, Vol. 8 (1960).

Malcolm, D. G. and A. J. Rowe, "Computer-Based Control Systems," *California Management Review*, Spring, 1961.

Malcolm, D. G., A. J. Rowe, and L. F. McConnell, eds., *Management Control Systems*. New York: John Wiley & Sons., Inc., 1960.

Mann, F. C. and R. L. Hoffman, *Automation and the Worker*. New York: Holt, Rinehart & Winston, Inc., 1960.

Mann, F. C. and L. K. Williams, "Observations on the Dynamics of a Change to Electronic Data-Processing Equipment," *Administrative Science Quarterly*, September, 1960.

Mangell, S., "PL—A New Concept in Complete Management Thinking," *Advanced Management*, August, 1960.

March, J. G. and H. A. Simon, *Organization*. New York: John Wiley & Sons, Inc., 1958.

Martin, E. W., "Simulation in Organizational Research," *Business Horizons*, Fall, 1959.

————, "The Widening Potential of the Computer," *Business Horizons*, Winter, 1958.

_____, ed., *Top Management Decision Simulation — The A.M.A. Approach*. New York: American Management Association, 1957.

Martin, J., *Design of Real-Time Computer Systems*. Englewood Cliffs, N.J.: Prentice-Hall, Inc., 1967.

_____, *Programming Real-Time Computer Systems*. Englewood Cliffs, N.J.: Prentice-Hall, Inc., 1965.

Martino, R. L., "A Generalized Plan for Developing and Installing a Management System," *Total Systems Letter*, April, 1965.

_____, "Balancing Production for Maximum Profit," *Total Systems Letter*, November, 1965.

_____, "Forecasting and the Systems Approach," *Total Systems Letter*, August, 1965.

_____, "Task Force Selection and Training," *Total Systems Letter*, July, 1965.

Marx, M. H., *Systems and Theories in Psychology*. New York: McGraw-Hill Book Company, 1963.

Mason, E. S., ed., *The Corporation in Modern Society*. Cambridge, Mass.: Harvard University Press, 1959.

Mattessich, R., "Budgeting Models and System Simulation," *The Accounting Review*, July, 1961.

Maxfield, M., A. Callahan and L. J. Fogel, *Biophysics and Cybernetic Systems*. Proceedings of the Second Cybernetic Sciences Symposium, Washington, D. C.: Sparton Books, Inc., 1965.

Mayo, H. R., "Putting the System in Writing," *Systems and Procedures Journal*, May, 1965.

Meistor, D. and G. F. Rabideau, *Human Factors Evaluation in Systems Development*. New York: John Wiley & Sons, Inc., 1965.

Meliz, P. W., "Impact of Electronic Data Processing on Managers," *Advanced Management*, April, 1961.

Merriam, C. W., *Optimization Theory and Design of Feedback Control Systems*. New York: McGraw-Hill Book Company, 1964.

Mesarovic, M. D., *Second Systems Symposium, Case Institute of Technology*. New York: John Wiley & Sons, Inc., 1964.

_____, *The Control of Multivariable Systems*. Cambridge, Mass.: Technology Press of the Massachusetts Institute of Technology and John Wiley & Sons, Inc., 1960.

_____, *Views on General Systems Theory*. Proceedings of the Second Systems Symposium, Case Institute of Technology, New York: John Wiley & Sons, Inc., 1964.

"Methodology of Modeling," *Electro-Technology*, September, 1966.

Meyer, H. A., ed., *Symposium on Monte Carlo Methods*. New York: John Wiley & Sons, Inc., 1956.

Michael, D. N., *Cybernation—The Silent Conquest*. Santa Barbara, Calif.: The Center for the Study of Democratic Institutions, 1962.

_____, "Social Environment," *Operations Research*, July, 1959.

Miller, D. W. and M. K. Starr, *Executive Decisions and Operations Research*. Englewood Cliffs, N.J.: Prentice-Hall, Inc., 1960.

Miller, J. J., "Automation, Job Creation, and Unemployment," *Academy of Management Journal*, December, 1964.

Miller, R. B., *A Method for Man-Machine Task Analysis*. Technical Report 53-136, Wright-Patterson Air Force Base, Ohio: Wright Air Development Center, 1953.

————, *Some Working Concepts of Systems Analysis*. Pittsburgh, Pa.: American Institute of Research, 1954.

Miller, R. B. and H. P. VanCott, *The Determination of Knowledge Content for Complex Man-Machine Jobs*. Pittsburgh, Pa.: American Institute of Research, 1955.

Miller, R. W., "How to Plan and Control with PERT," *Harvard Business Review*, March-April, 1962.

Mishkin, E., ed., *Adaptive Control Systems*. New York: McGraw-Hill Book Company, 1961.

Mohan, C., R. C. Garg, and P. P. Singal, "Dependability of a Complex System," *Operations Research*, Vol. 10 (1962).

Monroe, W. F., "Objectivity and Automated Management Decisions," *The Journal of Industrial Engineering*, March, 1966.

Moravec, A. F., "Basic Concepts for Designing a Fundamental Information System," *Management Services*, July-August, 1965.

————, "Basic Concepts for Planning Advanced Electronic Data Processing Systems," *Management Services*, May-June, 1965.

————, "Using Simulation to Design a Management Information System," *Management Services*, May-June, 1966.

Morgenthaler, G. W., "The Theory and Application of Simulation in Operations Research," in Ackoff, R. L., ed., *Progress in Operations Research*. New York: John Wiley & Sons, Inc., 1961.

Morris, W. T., *The Analysis of Management Decisions*. Homewood, Ill.: Richard D. Irwin, Inc., 1964.

Morse, G. E., "Pendulum of Management Control," *Harvard Business Review*, May, 1965.

Morse, P. M., *Queues, Inventories, and Maintenance*. New York: John Wiley & Sons, Inc., 1958.

Morse, P. M. and G. E. Kimball, *Methods of Operations Research*. New York: John Wiley & Sons, Inc., 1951.

Moss, J. H., "Notes on the Selection of a Computer for Simulation Purposes," in *Proceedings of the Seventh Conference on the Design of Experiments in Army Research Development and Testing*. Durham, N.C.: U. S. Army Research Office, ARODR 62-2, August, 1962.

Moss, R., "Human Engineering — Building Products to Fit People," *Management Review*, March, 1960.

Murphy, G. J., *Basic Automatic Control Theory*. Princeton, N.J.: D. Van Nostrand Publishing Co., 1950.

Murphy, G., "Toward a Field Theory of Communication," *Journal of Communication*, December, 1961.

Murrell, K. F., *Human Performance in Industry*. New York: Reinhold Publishing Corp., 1965.

Muschamp, G. M., "Tomorrow's Integrated Office and Plants," *Automation*, May, 1961.

Nadler, G., *Factors in Work Design*. Homewood, Ill.: Richard D. Irwin, Inc., 1963.

Nemhouser, G. L., *Introduction to Dynamic Programming*. New York: John Wiley & Sons, Inc., 1966.

Neuschel, R. F. *Management by System*. New York: McGraw-Hill Book Company, 1960.

Neuendorf, C. W., "The Total Management Information System," *Total Systems Letter*, March, 1965.

Neuschel, R. F., "Systems Man—Architect or Journeyman," *Systems and Procedures Journal*, January-February, 1963.

Newman, A. D., *Manual for GPSS I, A General Purpose Systems Simulator Designed by Geoffrey Gordon*. Armonk, N.Y.: Advanced Systems Development Laboratory, I.B.M. ASDD 17-077, May, 1962.

Newton, G. C., *Analytical Design of Linear Feedback Controls*. New York: John Wiley & Sons, Inc., 1957.

Noonan, G. C. and C. G. Fain, "Optimum Preventive Maintenance Policies When Immediate Detection of Failure Is Uncertain," *Operations Research*, Vol. 10 (1962).

O'Malley, R. L., S. E. Elmaghraby and J. W. Jeske, "An Operational System for Batch-Type Production," *Management Science*, June, 1966.

Operations Research Group, Case Institute of Technology, *A Comprehensive Bibliography on Operations Research, 1957-58*. New York: John Wiley & Sons, Inc., 1963.

Operations Research Group, Case Institute of Technology, *A Comprehensive Bibliography on Operations Research through 1956*. New York: John Wiley & Sons, Inc., 1958.

Optner, S. L., *Systems Analysis for Business Management*. Englewood Cliffs, N.J.: Prentice-Hall, Inc., 1960.

————, *Systems Analysis for Business and Industrial Problem Solving*. Englewood Cliffs, N.J.: Prentice-Hall, Inc., 1965.

Ornstein, G. N., *Simulation as a Tool in Evaluating System Effectiveness*. Los Angeles: North American-Rockwell, n.d.

Orr, D., "Two Books on Simulation in Economics and Business," *The Journal of Business*, June, 1963.

Papoulis, A., *Probability, Random Variables and Stochastic Processes*. New York: McGraw-Hill Book Company, 1965.

Pappo, H. A., *Theory of System Organization—I, Introduction to Control Systems.* System Development Corporation Report, SP-120, October, 1959.

Parker, R. W., "The SABRE System," *Datamation*, September, 1965.

Parsons, H. M., *System Trouble-Shooting.* System Development Corporation Report, SP-793, July, 1962.

Pask, G., *An Approach to Cybernetics.* New York: Harper & Row, Publishers, 1962.

Paynter, H. M., *Analysis and Design of Engineering Systems.* Cambridge, Mass.: The MIT Press, 1961.

Peschom, J., *Automatic Control.* Waltham, Mass.: Blaisdell Publishing Co., 1965.

———, *Disciplines and Techniques of Systems Control.* Waltham, Mass.: Blaisdell Publishing Co., 1965.

Peterson, E. L., *Statistical Analysis and Optimization of Systems.* New York: John Wiley & Sons, Inc., 1961.

Pfiffner, J. M. and F. P. Sherwood, *Administrative Organization.* Englewood Cliffs, N.J.: Prentice-Hall, Inc., 1960.

Pollock, N., "Determining the Authority and Scope of Systems Personnel," *Office*, May, 1959.

Popov, E. P., *Automatic Control.* Reading, Mass.: Addison-Wesley Publishing Co., Inc., 1962.

Porter, A., *An Introduction to Servomechanisms.* New York: John Wiley & Sons, Inc., 1950.

Porter, J. C., M. W. Sasieni, E. S. Marks and R. L. Ackoff, "The Use of Simulation as a Pedagogical Device," *Management Science*, February, 1966.

Postley, J. A., *Computers and People.* New York: McGraw-Hill Book Company, 1963.

Pritsker, A. B., "The Monte Carlo Approach to Setting Maintenance," *The Journal of Industrial Engineering*, May-June, 1963.

Program Planning and Control System. Washington, D. C.: Special Projects Office, Department of the Navy, 1960.

Pullen, K. A., *Design of a Communications System.* New York: Hayden Book Company, Inc., 1963.

Putnam, A. O. and E. R. Barlow et. al., *Unified Operations Management.* New York: McGraw-Hill Book Company, 1963.

Quade, E. S., *Military Systems Analysis.* Santa Monica, Calif.: The Rand Corporation, P-2676, November, 1962.

———, *Analysis for Military Decisions.* Chicago: Rand McNally & Co., 1964.

Raach, F. R., "Management Information Systems," *Duns Review and Modern Industry*, January, 1966.

Rabb, E. S., "Applied Psychology for the Systems Man," *Systems and Procedures Journal*, March, 1961.

Raben, M. W., *A Survey of Operations and Systems Research Literature.* Medford, Mass.: Tufts University, Institute for Applied Experimental Psychology, January, 1960.

Radamaker, T., *Business Systems.* Cleveland, Ohio: Systems and Procedures Association, 1963.

Radell, N. J., "Operations Research Techniques as a Basis for System Design," *Systems and Procedures Journal*, March, 1961.

Rader, L. T., "Roadblocks to Progress in the Management Sciences and Operations Research," *Management Science*, February, 1965.

Ragazzini and Franklin, *Sample-Data Control Systems*, New York: McGraw-Hill Book Company, 1965.

Ramo, S., "Weapon System Management," *California Management Review*, Fall, 1958.

Read, W. H., "The Decline of the Hierarchy in Industrial Organizations," *Business Horizons*, Fall, 1965.

Ream, N. J., "On-Line Management Information," *Datamation*, March, 1964.

Redfield, C. E., *Communication in Management.* Chicago: University of Chicago Press, 1958.

Redgrave, M. J., *Some Approaches to Simulation, Modeling, and Gaming at SDC.* System Development Corporation Report, SP-721, March, 1962.

Reid, P. C., "How to Survive a Systems Change," *Supervisory Management*, April, 1964.

Reintjex, J. E., "The Intellectual Foundations of Automation," *Annals of the American Academy of Political and Social Science*, March, 1962.

Reitzfeld, M., "Effective Reports: The Poor Man's Management Information System," *Records Management Journal*, Winter, 1964.

————, "Marketing the Systems Function," *Systems and Procedures Journal*, November, 1965.

Retterer, R. W., "Computers and Communications Provide New Conception in Scientific Management," *Advanced Management Journal*, October, 1964.

Reza, F. M., *An Introduction to Information Theory.* New York: McGraw-Hill Book Company, 1961.

Rhine, R. J., *Command-and-Control and Management Decision Making.* System Development Corporation Report, SP-1174, May, 1963.

Richard, M. D. and P. S. Greenlaw, *Management Decision Making.* Homewood, Ill.: Richard D. Irwin, Inc., 1966.

Roberts, E. B., "Industrial Dynamics and the Design of Management Control Systems," *Management Technology*, December, 1963.

Robinchaud, L. P. *et. al.*, *Signal Flow Graphs and Applications*. Englewood Cliffs, N.J.: Prentice-Hall, Inc., 1962.

Robins, W. R., "Theory and Design of the Management Information System," *Systems and Procedures Journal*, November-December, 1965.

Rogers, R. E., "Factored Cost—A Management Systems Approach to Cost Control," *Systems and Procedures Journal*, November-December, 1965.

Roman, D. D., "The PERT System," *Journal of the Academy of Management*, April, 1962.

Rose, T. G. and D. E. Farr, *Management Control*. New York: McGraw-Hill Book Company, 1957.

Rosenzweig, J., "The Weapon System Management and Electronic Data Processing," *Management Science*, January, 1960.

Rowe, A. J., "Research Problems in Management Controls," *Management Technology*, December, 1961.

Roy, H. J., "Info-Transfer System to Assist Managers in Planning and Controlling," *Controller*, October, 1959.

Ruesch, J. and G. Bateson, *Communication*. New York: W. W. Norton & Company, Inc., 1951.

Saaty, T. L., *Elements of Queuing Theory with Applications*. New York: McGraw-Hill Book Company, 1961.

Samuel, A. L., "Artificial Intelligence—A Frontier of Automation," *Annals of the American Academy of Political and Social Science*, March, 1962.

Sandler, G. H., *System Reliability Engineering*. Englewood Cliffs, N.J.: Prentice-Hall, Inc., 1963.

Sasieni, M., A. Yaspan and L. Friedman, *Operations Research — Methods and Problems*. New York: John Wiley & Sons, Inc., 1959.

Savant, C. J., *Basic Feedback Control System Design*. New York: McGraw-Hill Book Company, 1958.

————, *Control System Design*. New York: McGraw-Hill Book Company, 1964.

Sayer, J. S., J. E. Kelly and M. R. Walker, "Critical Path Planning," *Factory*, July, 1960.

Schlosser, R. E., "Psychology for the Systems Analyst," *Management Services*, November-December, 1964.

Schnorr, C. G., "Human Engineering," *Journal of Industrial Engineering*, November-December, 1958.

Schoderbek, P. P., *Management Systems, A Book of Readings*. New York: John Wiley & Sons, Inc., 1967.

————, "PERT/COST—Its Values and Limitations," *Management Review*, March, 1966.

Schoderbek, P. P. and C. G. Schoderbek, "The New Manager," *Systems and Procedures Journal*, July-August, 1965.

Schwartz, P. L., "Two Missing Links in the Systems Change," *Administrative Management*, July, 1965.

Schwartz, R. J. and B. Friedland, *Linear Systems*. New York: McGraw-Hill Book Company, 1965.

Schwitter, J. P., "Computer Effect Upon Management Jobs," *Journal of the Academy of Management*, September, 1965.

Scott, W. G., "Organizational Theory — An Overview and an Appraisal," *Journal of the Academy of Management*, April, 1961.

Seely, S., *Dynamic Systems Analysis*. New York: Reinhold Publishing Corp., 1964.

Seifert, W. W. and C. W. Steeg, *Control Systems Engineering*. New York: McGraw-Hill Book Company, 1960.

Sethi, N. K., "Management by System," *Personnel Journal*, April, 1963.

Self-Organizing Systems. Interdisciplinary Conference on Self-Organizing Systems, New York: Pergamon Press, Inc., 1960.

Shackel, B., "Ergonomics in Equipment Design," *Instrument Practice*, June, 1961.

Shenton, D. W. and H. Gleixner, "Automated Material Control," *Automation*, January, 1961.

Shera, J. H., ed., *Information Systems in Documentation*. New York: Interscience Publishers, 1957.

Shinners, S. M., *Control Systems Design*. New York: John Wiley & Sons, Inc., 1964.

Shober, J. A., "Decision Tables for Better Management," *Systems and Procedures Journal*, March, 1966.

Shubik, M., "Simulation of Industry and Firm," *American Economic Review*, December, 1960.

Schuchman, A., *Scientific Decision Making in Business*. New York: Holt, Rinehart & Winston, Inc., 1963.

Shultz, G. P. and T. L. Whisler, *Management Organization and the Computer*. Chicago: University of Chicago Press, 1960.

Shycon, H. N. and R. B. Maffei, "Simulation—Tool for Better Distribution," *Harvard Business Review*, November-December, 1960.

Simon, H. A., *Administrative Behavior*. New York: The Macmillan Company, 1959.

————, "Comments on the Theory of Organizations," *American Political Science Review*, December, 1952.

————, "How Computers Will Reshape the Management Team," *Steel*, January 11, 1965.

————, *Models of Man — Social and Rational*. New York: John Wiley & Sons, Inc., 1957.

————, *The New Science of Management Decision*. New York: Harper & Row Publishers, 1960.

————, *The Shape of Automation for Men and Management*. New York: Harper & Row, Publishers, 1965.

Simon, H. A. and A. Newell, "Heuristic Problem Solving—The Next Advance in Operations Research," *Operations Research*, January-February, 1958.

Simon, L. and R. Sisson, "Evolution of a Total System," *Total Systems Letter*, January, 1966.

Simonton, W. C., ed., *Information Retrieval Today*. Minneapolis, Minn.: University of Minnesota, Center for Continuation Study, 1963.

"Simulation—Shorter Time from Dream to Onstream," *Steel*, December 13, 1965.

Sinaiko, H. W., *Selected Papers on Human Factors in the Design and Use of Control Systems*. New York: Dover Publications, Inc., 1961.

Singer, C., *The Utilization of Computers and Computerized Systems in Management*, System Development Corporation Report, SP-493, May, 1962.

Smith, D. B., "Systems Engineering—Implications for Management," *Financial Analysis Journal*, May, 1965.

Smith, J. K., "Chrysler Instantaneous Quality Reporting," *Industrial Quality Control*, January, 1967.

Smith, K. M., *A Practical Guide to Network Planning*. London: British Institute of Management, 1965.

Smith, K. U. and M. F. Smith, *Cybernetic Principles of Learning and Educational Design*. New York: Holt, Rinehart & Winston, Inc., 1966.

Smith, O. J., *Feedback Control Systems*. New York: McGraw-Hill Book Company, 1958.

Solodovnikov, V. V., *Introduction to the Statistical Dynamics of Automatic Control Systems*. New York: Dover Publications, Inc., 1960.

Spaulding, A. T., "Is the Total System Concept Practical?" *Systems and Procedures Journal*, January, 1964.

Spencer, M. H., "Computer Models and Simulation in Business and Economics," *Business Topics*, Winter, 1963.

————, "The Framework of Management Decision Making," *Business Topics*, Summer, 1963.

Sprague, R. E., *Electronic Business Systems*. New York: The Ronald Press, Inc., 1962.

————, "Electronic Business Systems — Nineteen Eighty-Four," *Business Automation*, February, 1966.

————, "On Line-Real Time Systems as a Long-Range Planning Goal," *Total Systems Letter*, April, 1966.

Starr, M. K., "Modular Production—A New Concept," *Harvard Business Review*, November-December, 1965.

————, *Production Management, Systems and Synthesis*. Englewood Cliffs, N.J.: Prentice-Hall, Inc., 1964.

Stern, T. E., *Theory of Nonlinear Networks and Systems—An Introduction*. Reading, Mass.: Addison-Wesley Publishing Co., Inc., 1965.

Stilian, G. M., "Impact of Automation on the Manufacturing Executive's Job," *Management Review*, March, 1958.

Stoller and VanHorn, "Design of a Management Information System," *Management Technology*, January, 1960.

Stout, T. M., "Process Control — Past, Present and Future," *Annals of the American Academy of Political and Social Science*, March, 1962.

"Systems Analysts Move into Managing," *Iron Age*, August 26, 1965.

Systems Contracting. New York: American Management Association, Purchasing Division, Management Bulletin No. 63, 1965.

"Systems for Total Management," *Data Processing for Management*, August, 1963.

Systems Planning and Control. New York: American Management Association, Special Report 12, 1956.

Svenson, A. L., "Congruency — Driving Force in Management Systems," *Systems and Procedures Journal*, May-June, 1966.

————, "Management Systems and the Exception Principle," *Systems and Procedures Journal*, July-August, 1964.

Tannenbaum, R., "Managerial Decision Making," *Journal of Business*, January, 1950.

Taube, M., *Computers and Common Sense*. New York: Columbia University Press, 1962.

Thaler, G. J. and R. G. Brown, *Analysis and Design of Feedback Control Systems*. New York: McGraw-Hill Book Company, 1960.

————, *Servomechanism Analysis*. New York: McGraw-Hill Book Company, 1953.

Thayer, L. M., "Hueristic Programming Techniques in Decision Making Systems," *Data Processing Yearbook—1965*, Detroit, Mich.: American Data Processing, Inc., 1964.

Thompson, H. A., *Joint Man-Machine Decisions—The Phase Beyond Data Processing and Operations Research*. Cleveland, Ohio: Systems and Procedures Association, 1965.

Thompson, J. D. and F. L. Bates, "Technology, Organization and Administration," *Administrative Science Quarterly*, December, 1957.

Thompson, V. A., "Hierarchy, Specialization and Organizational Conflict," *Administrative Science Quarterly*, March, 1961.

Thurston, P. H., *Systems and Procedures Responsibility*. Boston, Mass.: Houghton Mifflin Company, 1959.

————, "The Concept of a Production System," *Harvard Business Review*, November, 1963.

————, "Who Should Control Information Systems?" *Harvard Business Review*, November-December, 1962.

Tilles, S., "The Manager's Job—A Systems Approach," *Harvard Business Review*, January-February, 1963.

Tocher, K. D., "Review of Simulation Languages," *Operations Research*, June, 1965.

Tomeski, E., "Needed — Management Systems Research," *Systems and Procedures Journal*, March-April, 1964.

Tomvic, R., *Sensitivity Analysis of Dynamic Systems*. New York: McGraw-Hill Book Company, 1963.

Tonge, F. M., "Summary of a Heuristic Line Balancing Procedure," *Management Science*, October, 1960.

————, "The Use of Heuristic Programming in Management Science," *Management Science*, April, 1961.

Tou, T. L., *Automatic Control*. New York: McGraw-Hill Book Company, 1959.

————, *Optimum Design of Digital Control Systems*. New York: Academic Press, Inc., 1963.

"Trends in Corporate Data Systems," *EDP Analyzer*, August, 1966.

Trudeau, A. C., "Modern Systems Management," *Edison Electric Industry Bulletin*, January, 1966.

Trull, S. G., "Some Factors Involved in Determining Total Decision Success," *Management Science*, February, 1966.

Tsien, H. S., *Engineering Cybernetics*. New York: McGraw-Hill Book Company, 1954.

Tucker, G. K. and D. M. Wills, *A Simplified Technique of Control System Engineering*. Philadelphia, Pa.: Minneapolis-Honeywell Regulator Company, 1958.

————, *Automatic Control*. Philadelphia, Pa.: Minneapolis-Honeywell Regulator Company, 1958.

Tuthill, O. W., "The Thrust of Information Technology on Management," *Financial Executive*, January, 1966.

Vazsonyi, A., "Automated Information Systems in Planning, Control and Command," *Management Science*, February, 1965.

————, *Scientific Programming in Business and Industry*. New York: John Wiley & Sons, Inc., 1958.

Vickery, B. C., *On Retrieval System Theory*. London: Butterworths, 1965.

VonBertalanffy, L., "An Outline of General Systems Theory," *British Journal of Philosophical Sciences*, (1950).

————, "General Systems Theory—A Critical Review," *Yearbook of the Society for General Systems Research*, Vol. 7 (1962).

————, "General System Theory," *General Systems*, Vol. 1 (1956).

————, "General System Theory — A New Approach to Unity of Science," *Human Biology*, December, 1951.

VonNeumann, J., *The Computer and the Brain*. New Haven, Conn.: Yale University Press, 1959.

Wegner, P., *Introduction to System Programming*. New York: Academic Press, Inc., 1962.

Weinberg, G. M., "Lifting the Veil from Systems Engineering," *Factory*, April, 1966.

Weinwurm, G. F., "Computer Management Control Systems Through the Looking Glass," *Management Science*, July, 1961.

Wescott, J. J., ed., *Exposition of Adaptive Control*. Proceedings of a Symposium Held at Imperial College of Science and Technology (London), New York: The Macmillan Company, 1962.

Wheatley, P. E., "Human Element in a Systems Survey," *Systems and Procedures Journal*, May, 1960.

Whisler, T. L., "Manager and the Computer," *Journal of Accounting*, January, 1965.

Wiener, N., *Cybernetics — Control and Communication in the Animal and the Machine*. New York: John Wiley & Sons, Inc., 1948.

————, *Cybernetics — Or Control and Communication in the Animal and the Machine*. Cambridge, Mass.: The MIT Press, 1961.

————, *God and Golem, Inc., — A Comment on Certain Points Where Cybernetics Impinges on Religion*. Cambridge, Mass.: The MIT Press, 1962.

————, *The Human Use of Human Beings — Cybernetics and Sociology*. Boston, Mass.: Houghton Mifflin Company, 1950.

————, *The Human Use of Human Beings*. Boston, Mass.: Houghton Mifflin Company, 1954.

Wilde, D. J., "Production Planning of Large Systems," *Chemical Engineering Progress*, January, 1963.

Williams, H. L., "Reliability Evaluation of Human Component in Man-Machine Systems," *Electric Manufacturer*, April, 1958.

Williams, L. K., "The Human Side of Change," *Systems and Procedures Journal*, July-August, 1964.

Williams, T. G., *The Design of Survivable Communications Networks*. Utica, N.Y.: Equipment Laboratory, Rome Air Development Center, RADC-TDR-62-403, October, 1962.

Williams, T. J., *Systems Engineering for the Process Industries*. New York: McGraw-Hill Book Company, 1961.

————, "Systems Engineering," *Chemical Engineering*, February, 1960.

Willmorth, N. E. and C. J. Shaw, *System Documentation*. System Development Corporation Report, SP-974, October, 1962.

Wilson, W. E., *Concepts of Engineering System Design*. New York: McGraw-Hill Book Company, 1965.

Wilson, I. G. and M. E. Wilson, *Information, Computers and System Design*. New York: John Wiley & Sons, Inc., 1965.

Wohlstetter, A. J., *Systems Analysis Versus System Design*. Santa Monica, Calif.: The Rand Corporation, P-1530 (1958).

Woodgate, H. S., *Planning by Network*. London: Business Publications, 1964.

Woods, R. S., "Some Dimensions of Integrated Systems," *The Accounting Review*, July, 1964.

Woodson, W. E. and D. W. Conover, *Human Engineering Guide for Equipment Designers*. Berkeley, Calif.: University of California Press, 1964.

Woolridge, D. E., "Operations Research — The Scientists Invasion of the Business World," in A. Schuchman, *Scientific Decision Making in Business*. New York: Holt, Rinehart & Winston, Inc., 1963.

Yovits, M. C. and S. Cameron, *Self-Organizing Systems*, New York: Symposium Publications Division, Pergamon Press, Inc., 1960.

Yovits, M. C., G. T. Jacobi and G. D. Goldstein, *Self-Organizing Systems*. Washington, D. C.: Sparton Books, 1962.

Young, S., *Designing a Behavioral System*. Boston: Proceedings of the 23rd Annual Meeting of the Academy of Management, December, 1963.

————, *Management: A Systems Analysis*. Glenview, Ill.: Scott Foresman and Company, 1966.

Ziessow, B. W., "Management by Exception," *Data Processing Magazine*, October, 1965.

Zolad, R. J., "Systems, Computers and Operations Research — One Company's Experience," *Systems and Procedures Journal*, September, 1965.

Index

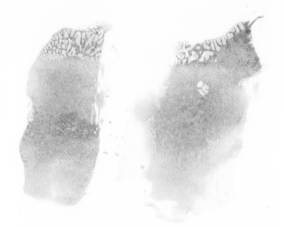